A WELFARE STATE STRIKES OIL

The Norwegian Experience

Walter Galenson

UNIVERSITY PRESS OF AMERICA

LANHAM • NEW YORK • LONDON

University Press of America,® Inc.

4720 Boston Way
Lanham, MD 20706

3 Henrietta Street
London WC2E 8LU England

Printed in the United States of America

Library of Congress Cataloging in Publication Data

Galenson, Walter, 1914-
A welfare state strikes oil.

Bibliography: p.
Includes index.
1. Norway—Economic policy. I. Title.
HC365.G33 1986 338.9481 86-1590
ISBN 0-8191-5273-0 (alk. paper)
ISBN 0-8191-5274-9 (pbk. : alk . paper)

All University Press of America books are produced on acid-free paper which exceeds the minimum standards set by the National Historical Publications and Records Commission.

CONTENTS

LIST OF TABLES

PREFACE

Many years ago I wrote a book entitled Labor in Norway in which I set forth the contributions of Norwegian management and labor in establishing an excellent system of industrial relations. For the two decades following the publication of that book, sound, cautious economic policies pursued by the government, mainly under the control of the Labor Party, led to a steady and substantial rate of economic growth that made Norway one of the richest countries in the world.

The discovery of oil on the Norwegian North Sea shelf created the opportunity of combining the advanced welfare state that was already in place by the early 1970s with a dynamic and smoothly operating economy, and of providing a model of a new social order that other nations might seek to emulate. This did not happen. The Labor government that was in power for most of the following decade expanded the welfare state, but in so doing it impaired the country's manufacturing sector and its export potential. It is the purpose of this book to explain how it happened.

A number of people were kind enough to read the manuscript and to give me the benefit of their advice. I refrain from mentioning their names only because I do not want to implicate them in my interpretations and conclusions. However, I do want to record my great obligation to Haakon Lie, who was General Secretary of the Norwegian Labor Party from 1945 to 1969. He urged me to undertake this study, and it is largely through him that, over the years, I gained some understanding of the politics and economics of the beautiful country that is the subject of this book.

January, 1986 Walter Galenson

CHAPTER I

THE NORWEGIAN ECONOMY ON THE EVE OF THE OIL DISCOVERIES

Norway entered the ranks of oil-rich nations in the 1970s. Extraction concessions for the Norwegian continental shelf were first awarded in 1965; within a decade, 100 holes had been drilled to search for deposits and to examine the extent of discoveries. Although there was some production in 1971, it was not until 1975 that there was a net export of oil. The impact of the oil find on the Norwegian economy thus dates from the mid-1970s.

Norway is not the only country to have attained sudden affluence through the discovery of oil or another natural resource, but it was one of the few that was already wealthy when the find occurred. In 1970, Norway ranked seventh in per capita gross domestic product (GDP) among the OECD nations. It was already considerably ahead of the United Kingdom, and close to the United States and Germany. Only Sweden and Switzerland were significantly richer by this measure.[1]

Before World War II Norway had been far from a wealthy country, and it had suffered severe economic loss from the German wartime occupation. Sound economic policies during the first two postwar decades resulted in an average annual real GDP increase of 4 percent from 1952 to 1970. Particularly impressive was the record of growth from 1959 to 1967, averaging 5.1 percent per annum, the highest registered for any eight-year period in Norwegian history. For a variety of reasons, the rate of growth fell from 1967 to 1970, but the earlier pattern resumed with a solid 4.6 percent GDP increase in 1971. The country seemed to be back on its earlier growth track.

For most of the postwar years, government was in the hands of the Norwegian Labor Party. It won a parliamentary majority in 1945 (there had been a period of Labor government before the war), and remained in power for the next twenty years, winning four general elections. The Norwegian parliament, which is called the Storting, is elected every four years, but unlike the practice of most other parliamentary bodies, members of the Storting serve a fixed four-year term. Substitutes replace members who resign or retire within the term period. A political party with an absolute majority is assured of a four-year tenure apart from the unlikely contingency of defection by members of its parliamentary group. A coalition can fall within the fixed term if the partners fail to agree on policy and bring about a vote of non-confidence.

The Labor Party lost its majority in 1961 and until 1965, governed with the support of a leftist socialist party. It was succeeded in 1965 by a non-socialist coalition. This fell apart in 1971, and a minority Labor government took over, pledged to full support of Norwegian entrance into the Common Market.

This issue polarized the country. The Conservative Party, representative of industrial and commercial interests, favored joining the Common Market, but its former coalition partners were opposed, as was a vociferous left wing in the Labor Party itself. A national referendum held in the fall of 1972 resulted in a 53 percent majority against the Market. In consequence, the Labor Party left the government, only to take over again after winning an election a year later.

During the twenty-eight year period from 1945 to 1973, the Labor Party governed Norway for twenty-one years. It is fair to say that it was the principal architect of economic policy during this period, particularly since there were some serious policy differences within the non-Labor coalitions that held office during the remaining seven years.

In its formative stage, the Labor Party was on the radical side of the European socialist spectrum. For a short time after World War I, it was affiliated with the newly formed Comintern. However, it refused to accept the dogmatic requirements for membership in that organization laid down by the Soviet leadership, and was expelled. It evolved into a moderate social democratic party, committed to raising the living standards of lower income groups through social welfare programs.

The OECD described Norway's economic performance prior to the 1970s in the following terms:

> Norway has for many years past demonstrated that it is possible to pursue social and egalitarian objectives in economic policy while maintaining high employment, reasonable price stability, external balance and a satisfactory rate of economic growth. For most of the 1950s and 1960s domestic cost and price developments in Norway can be

> shown to have been largely determined by international prices and the growth rate of labor productivity in export and import-competing industries.[2]

The prosperity that Norway had attained in 1970 was based upon a productive and varied industry. In that year, manufacturing accounted for 25 percent of the domestic product while merchant shipping accounted for another 11 percent.

As a small country, Norway was of necessity heavily involved in foreign trade. Some 40 percent of the total output of its goods and services was exported in 1970. Norway was Europe's largest exporter of cement. Based upon cheap electricity from well developed water power resources, it was by far the largest producer of aluminum in Western Europe, and one of the two or three largest aluminum exporters in the world. Other major exports were pulp and paper, based upon its forest resources, and chemical fertilizers, another energy intensive industry. Other major exports included ships, machinery and equipment, and fish products.

Norway normally ran commodity trade deficits. In 1970, for example, imports exceeded exports by 52 percent. The gap was largely filled by earnings of the merchant marine, and foreign borrowing was minimal. But the economy was vulnerable to business conditions abroad; a decline in shipping or a loss of foreign markets for traditional exports could cause serious problems. The key to Norway's ability to maintain a relatively favorable foreign balance was the cost competitiveness of its export industries. An average level of productivity performance was compensated for by moderate increases in wages.

Pre-oil age affluence owed a great deal to a sound exploitation of Norway's natural resources, chief among them cheap hydro-electric power. Domestic sources of energy covered two-thirds of domestic requirements. The price of electricity was kept low, thus encouraging the expansion of energy-intensive industries. The large forested areas of the country provided the base for prosperous lumber, pulp, and paper industries. By the early 1970s, however, the supply of raw materials was falling behind demand, necessitating raw material imports from Sweden.

A costly policy error attributable to the Labor Party

was the establishment of a state-owned basic steel industry after World War II. It was believed at the time that there would be a world shortage of steel, and that in any event, cheap energy would make the industry competitive. This expectation did not materialize, and the enterprise proved to be a constant drain on resources.

The agricultural sector has been a continual economic problem. Norway produces about half the amount of grain it uses, and has been a net exporter of proteins due to its fishing industry. The country is about 40 percent self-sufficient in food, calculated in energy content. But farms are small and uneconomic and can survive only with heavy subsidization. For some years, the combination of income compensation and food subsidies raised state financial support to the farm sector to more than 100 percent of factor income, so that the national income might have been higher with no farming at all. As in many countries, farming is a political sacred cow in Norway, and a great deal of ingenuity has gone into the development of a rationale for keeping it going at its current levels.

In the past, Norway benefited from a system of collective bargaining that operated quite well, on the whole. A so-called Contact Committee, consisting of members of the government and representatives of employers, employees, farmers, and fishermen, served as a forum for exchange of views on wage and price developments. A Technical Expert Committee, which also included representatives of the interest groups, provided advice on the economic background to negotiations. The formal wage agreements between the trade unions and employer associations were generally moderate and consistent with economic stability.

The system's major blemish was the practice of conducting local plant bargaining after national agreements had been signed. The result was extra-contract wage drift that sometimes exceeded the contract increases, a problem that was to worsen in the 1970s. Nevertheless, it was estimated that for the period 1961 to 1970, labor costs, including non-wage costs, rose by 8.4 percent per annum on average, compared with 5.5 percent for productivity, so that only moderate price increases were necessary.[3] The OECD summarized Norway's wage-price experience during the 1960s as follows:

> The merits of the Norwegian incomes policy, besides being an investment

> in income distribution, lies in the fact that it has enabled the government to keep the use of productive resources on a very high level throughout the postwar years. While the average increase in prices and production costs does not differ much from developments in most other countries, unemployment has usually been lower than the European average, and economic growth has been rapid and steady.[4]

How well off were the Norwegian people before the advent of oil? The answer is, very well off by international standards. World Bank data suggest that living standards in 1970 Norway were well above the average of those prevailing in the developed nations.

Greater income equality has long been a major objective of the Labor Party. Its success in achieving this goal can be judged from the various indices of income distribution shown in Table I-1 (the lower the index figures, the less the income inequality). While reliable income distribution data are not available for some countries, it is likely that Norway, together with Sweden, stood at the world's apex of income equality *circa* 1970.

Table I-1. Alternative Measures of Income Inequality, Norway and Selected Countries

	Gini	Kuznets	Theil
Norway (1970)	0.307	0.247	0.067
Sweden (1972)	0.302	0.242	0.064
Netherlands (1967)	0.354	0.284	0.091
United Kingdom (1973)	0.318	0.252	0.072
Germany (1973)	0.383	0.314	0.108
United States (1972)	0.381	0.306	0.204

Source: Malcolm Sawyer, *Income Distribution in OECD Countries*, OECD Economic Outlook, Occasional Studies, July, 1976, p. 17.

On the eve of its entrance into the ranks of the major oil producers, Norway's economy had reached an enviable level. High living standards prevailed, poverty had all but been eliminated. There was a varied and profitable industry that could compete internationally.

The Labor Party, which had governed the country during most of the post-World War II years, seemed to be pragmatic in its economic outlook, more so than most of the European socialist parties. A close alliance between the Labor Party and the trade unions made for incomes policy arrangements that worked reasonably well.

This is not to say that the economy operated with textbook efficiency. Large farm subsidies and dubious government investments in industry were dark spots. The level of taxation needed to support the burgeoning welfare state was already among the highest in the world. There were varying views about appropriate economic policy among the various interest groups, but they were not as sharply divisive and troublesome to domestic tranquility as in most other countries. The outlook was for a continuation of steady economic growth and gradual improvement of an already favorable situation.

CHAPTER II

OIL

The Resources

The first major discovery in Norwegian waters occurred in 1969, when the Phillips company struck oil in the Ekofisk field. This was the only producer until 1977, when the Friggfield came into operation, followed by Statfjord two years later. The first significant amounts were produced in 1978, and thereafter output mounted rapidly. Annual production figures appear in Table II-1.

Table II-1. Norwegian Crude Oil and Natural Gas Output, 1971-1984

(million ton equivalents)[a]

Year	Total	Oil	Gas
1971	0.3	0.3	---
1972	1.6	1.6	---
1973	1.6	1.6	---
1974	1.7	1.7	---
1975	8.7	8.7	---
1976	13.6	13.6	---
1977	16.7	13.6	3.1
1978	31.9	17.0	14.9
1979	40.4	18.8	21.6
1980	50.5	24.5	26.0
1981	49.7	23.5	26.2
1982	50.7	24.5	26.2
1983	56.3	30.5	25.8
1984	62.5	35.0	27.5

Source: Central Bureau of Statistics of Norway, Statistical Yearbook 1985, p. 173

[a]One ton equivalent is equal to one ton of oil and about 1,000 cubic metres of gas.

By 1984, some 386 million t.o.e. had been recovered from fields on the Norwegian continental shelf. The remaining recoverable reserves in the developed fields were estimated at 900 million t.o.e., 60 percent of which is oil and 40 percent gas. There are about 1500 million t.o.e. in fields not yet developed but thoroughly mapped.

In addition to these reserves, the Petroleum Directorate estimates that there are an additional 2000 million t.o.e. of recoverable reserves south of the 62nd parallel, of which 1300 million t.o.e. are low-risk, i.e., are likely to yield the estimated amounts, and 700 million t.o.e. are of variable risk.

The area north of the 62nd parallel is known to contain a high hydrocarbon potential, but estimates of recoverable oil vary enormously, from 2,000 to 17,000 million t.o.e. Whatever the final result, the reserves are substantial and will last for a considerable number of years at current rates of production.[1]

Norway itself consumes about 7 million t.o.e. a year, so that the bulk of the output is available for export. Although its current output is only about one percent of total world production, Norway is a significant supplier to Western Europe. In 1981, about 5 percent of oil imports and 45 percent of gas imports to Western Europe came from Norway, and these figures will rise as Norwegian production increases. Norway is no Saudi Arabia -- its proven reserves are perhaps only 10 percent as large as the Saudi's -- but it is a producer of some significance even at the present time.

Economic Importance of Oil

Even with the limited production thus far attained, oil and gas have had a major impact on the Norwegian economy. Table II-2 presents some of the indicators by which this impact can be measured. It tells a dramatic story. Over a decade, the contribution of the oil sector to the national product rose from zero to 17 percent, and to a third of exports. For 1983, the oil sector absorbed 19 percent of total investment and provided almost a sixth of government revenue.

For some years of the decade, the entire growth of the GDP was generated by the oil sector. A special feature of this sector was the very large profit (or rent) it yielded, estimated at almost 75 percent of the gross value of production. Most of this found its way into government coffers in one form or another. OPEC deserves a great deal of the credit for this fortuitous development. The Norwegian government did not determine prices, which typically exceeded the average OPEC level because of the high quality of North Sea oil.

Table II-2. Oil and Gas in the Norwegian Economy, 1970-1983

(oil and gas as percent of stated magnitudes)

	Gross Domestic Product	Gross Investment	Total Exports	Government Revenues
1970	----	1.5	----	----
1971	----	2.7	----	----
1972	0.2	4.5	0.5	0.1
1973	0.2	9.5	0.6	0.2
1974	0.8	14.2	1.3	0.3
1975	2.9	15.1	6.7	2.2
1976	4.1	17.5	10.8	4.6
1977	4.5	19.2	12.5	4.4
1978	7.0	14.4	18.1	6.9
1979	9.9	10.6	23.6	9.9
1980	15.7	9.0	33.2	18.2
1981	17.0	19.3	33.8	18.4
1982	17.0	15.1	35.3	18.8
1983[a]	17.0	19.0	33.3	16.0

Source: Norges Offentlige Utredninger, Petroleumsvirksomhetens framtid, 1983, No. 27, p. 18; Ministry of Petroleum and Energy, Report No .32 to the Storting, 1984-85, pp. 55-56.

[a]Preliminary estimates

The effect of oil on Norwegian business conditions was summarized by the Central Bureau of Statistics:

> The effects are manifested in several ways. Changes in the volume of investments on the Norwegian shelf are reflected in new orders and production in manufacturing industry. Changes in the rate of production and prices of oil and gas have an immediate effect on the figures for the domestic product and exports. The use of oil revenue domestically has spin-off effects that influence most aspects of our economy. In sum, oil activities mean that what we hitherto regarded as the traditional cyclical pattern, *i.e.*,

> that developments abroad are transmitted to the Norwegian economy through exports of traditional manufactured goods, has gradually become less important.[2]

North Sea oil is a particularly capital-intensive industry because of natural conditions at the production sites. Heavy investment in drilling and production platforms is required before any oil is recovered. However, the impact of investment in oil upon available investment resources was minimized by the fact that until 1980, at least, 90 percent of capital requirements were financed by foreign banks and oil companies, while the rest was provided by Norwegian oil firms.[3]

The direct impact of the oil industry on employment has not been great. Some 52,000 people were employed in this sector in 1983, about 3 percent of total employment. About 15,000 of these were engaged in exploration, boring, and production, while the rest were in catering and transportation to the oil platforms and in building platforms and supply ships. The indirect effects on the labor market have been substantial, however, and they will be discussed below. Suffice it to note here that a 1982 study indicated that half the enterprises that lost personnel to the oil industry had difficulty recruiting replacements, and that two-thirds of enterprises and institutions blamed their personnel shortages on oil. Civil engineers, welders, lathers, and carpenters were in particularly short supply. The employment activity of the oil sector has given rise to what is called a "vacuum cleaner effect," pulling skilled labor from all sectors of the economy in what was already a tight labor market.[4]

The Administrative Structure

Given the socialist ideology of the Labor Party, it was inevitable that the government should play a major role in managing the oil industry. The State Oil Company, known as Statoil, was established in 1972, with the following mandate:

> Statoil will be an important institution for implementing the petroleum policy of government authorities. Guidelines for Statoil's functions will be decided by the relevant political institutions. . . . Among

> other functions, this enterprise will be entrusted with the task of acting as an independent operator as soon as possible. This means that the company will be responsible for technical exploration and extraction of petroleum.[5]

At the same time, the government set up a Petroleum Directorate, later subordinated to the Ministry of Oil and Energy, to monitor political control of Statoil.

The earliest round of exploration concessions did not involve any state participation, but after its creation, Statoil was awarded a 50 percent participation in all subsequent concessions. Other Norwegian firms, notably Norsk Hydro, the large chemical combine, were assigned operational duties. As time went on, the Norwegian share both of proprietary rights and operations tended to rise. In 1981, for example, half the exploratory drilling was done by Norwegian companies, which owned two-thirds of new blocks.

Despite the buildup of their competence, Norwegian companies do not yet have the technical capacity to carry out complete development of an entire field, in part because of a shortage of qualified personnel. The government conceded in 1982 that "a definite need for the expertise of foreign companies in exploration and extraction will continue. . . . It will be necessary to assess the need for foreign assitance from project to project."[6]

Statoil advises the government in allocating concessions and is the principal collector of government oil revenue. With 2600 employees, it is in a position to oversee the activities of all other participating enterprises, both foreign and domestic.

This administrative structure was criticized by an outside agency after a survey conducted in 1981. The dominant role of Statoil, the report stated, did "not provide an environment in which the foreign and domestic oil companies will freely express their technical opinions. The status of Statoil made it difficult to determine whether it was run efficiently. Nor did the government have adequate control over Statoil because of inadequate staffing in the responsible ministry." The government was urged to transform Statoil into a "normal" company with a "normal" debt/equity structure and then to monitor its activities through an annual budget.[7]

Changing the administrative structure raises delicate political problems. There is no sign that the nonsocialist government plans to dismantle Statoil. However, in selling oil leases in 1984, more drilling blocks were released to the international oil companies than in any sale since the first one in 1963.

Early Policy Decisions

When it became evident in the early 1970s that oil was destined to play a major role in Norway's future, the Ministry of Finance issued a comprehensive report in which some guidelines for government action were set out. A year later, a second report which dealt with other natural resources as well as oil, which amplified the earlier document, was published.[8]

In the early euphoria of the oil discovery, it was estimated that by 1977, output would reach the "moderate" level of 60 million t.o.e., increasing to 90 million tons by 1981-82. The belief was that the latter level could be maintained at least until the end of the century. Several political parties argued for lower targets, but both Labor and Conservatives considered the 90 million ton target to be realistic, and this goal was confirmed by the Storting.

Report No. 25 emphasized that there should be "democratic control" of the oil. It was cautious on the use of oil revenues for immediate increases in consumption and urged that future generations be taken into account. However, it did see new possibilities for greater investment in social infrastructure and for expansion of the welfare state. Leveling regional and class income differentials was set as a major objective. Various ways of utilizing oil revenues for foreign investment were outlined. It was recognized that there might be some adverse effects, so that "the direct welfare advantages, such as better job opportunities, eased taxation, improved public contributions, etc., must be weighed against undesirable readjustments in employment and changes in residence."[9] The Report was interesting and remarkably accurate in some of its predictions.

Report No. 50 elaborated on some of these themes. Improvement of health services, expansion of kindergartens and universities, better sewage systems and housing conditions, higher pensions, all were to be made possible when the oil began to flow. A 17 percent growth of public employment was foreseen by 1980. A golden future was opening up, with more of everything for everyone.

Some cautionary voices were raised. Knut Getz Wold, governor of the central bank, urged that a substantial portion of the future revenues should be invested abroad. He warned that agricultural, fishing, and traditional manufacturing would face stiff competition from oil related industry. Much of the increased consumption would come in the form of goods and services produced by labor-intensive industries, putting greater pressure on the labor market. "When hotel and restaurant prices reach prohibitive levels in Norway, that trip abroad is more tempting." Export industries would become less competitive due to higher labor costs.[10]

Per Kleppe, who was minister of finance in the Labor government and the principal architect of Labor's economic policy during the 1970s, was more sanguine. He saw nothing wrong with borrowing abroad in anticipation of oil revenues to finance industrial subsidies, to create jobs, to raise unemployment benefits, to expand training facilities, etc. ". . . even without the prospective oil income Norway should pursue more or less the same policy, viz, to maintain activity and employment in spite of the foreign recession. . . . Foreign borrowing was an essential prerequisite for a high level of investment and provided the basis for a higher standard of living in the longer term."[11]

In any event, the "moderate" forecast of 90 million tons by 1980 proved far too optimistic. Technical difficulties and delays in field development were responsible for the slower growth of production, although the shortfall was partially made up by the OPEC price increases. Work injuries were not infrequent, and the destruction of the Alexander Kielland oil platform, with a considerable loss of life, provided a grim reminder that oil recovery was a hazardous enterprise.

Perhaps the most damaging effect of early oil optimism was the creation of great expectations of immediate affluence among the Norwegian people. When some of these expectations were not realized, an economist remarked: "Do let us remember that we found oil in the North Sea, not nurses." He might have added homes for the aged, roads, nurseries, and many other potential beneficiaries of state and private expenditures.

The issue to be pursued is not whether the oil discoveries led to improved living standards and a rise in the general level of welfare; of course they did. Rather, the question is whether the policies pursued by the various governments that held power after 1973, particularly Labor, used the oil revenues in an economically efficient manner, and whether they took sufficient account of the impact of the oil sector on the rest of the economy. Would the present Norwegian economy have been in a stronger position if a different set of policies had been followed? These are much more difficult questions to answer.

CHAPTER III

LABOR'S POLICY OBJECTIVES

Norway was already a full fledged welfare state by the beginning of the 1970s. In common with its Scandinavian neighbors, the government taxed away a large proportion of the GDP and redistributed it on the basis of a number of criteria, only some of which involved income levels. It has been observed that in Scandinavia, there has been "a greater willingness than [in] most other countries to use the public authorities -- state as well as regional and local -- as a tool to serve certain political targets such as security, the elimination of poverty, and equality in income and power. . . . No one is denied help when it is needed because of occupational status, sex, age, or low income."[1]

In the normal course of events, Labor would probably have moved cautiously to broaden the scope of the welfare system and to raise the benefit levels until checked by budget deficits and balance of payments constraints, as Denmark and Sweden were obliged to do. Oil revenues removed these constraints. The trend toward greater government intervention and spending was reinforced by the parliamentary elections of 1973, in which the Labor Party made its worst showing since the 1930s.

The Labor Party had come out in support of Norwegian entrance into the Common Market, but a left wing minority group within it joined with farmers and other social groups to bring about a negative vote in a referendum held in 1972. Labor Prime Minister Trygve Bratteli resigned, and before the 1973 elections were held, a new left alliance was formed, which gained 16 seats in the Storting, while Labor's share fell from 74 to 62. However, the Labor Party and the leftists together retained a small parliamentary majority, and the Labor Party formed a minority government that was hostage to a grouping of assorted Marxists who set themselves up as the Left Socialist Party. Although this organization did not last long, partly because of the unwillingness of the communists to merge their identity with the non-communist (and often more radical) left, its Storting representation served to push the Labor Party to the left.

The 1977 elections saw an increase in Labor Party representation to 76 seats, two short of a majority, while the left socialists were down to two seats. The non-socialist parties might have attempted to govern as a minority, but they were divided on a number of issues, including abortion and the pace of oil development. Another Labor minority government was formed which could have been overthrown by adverse votes of the two left socialists.

There was a great deal of conflict within the Labor Party during these years between those who favored continuing the moderate, pragmatic policies that had prevailed during the previous two decades, and members who, either out of conviction or to attract left-wing support, called for a more radical line. Odvar Nordli, a centrist, replaced the more conservative Trygve Bratteli, who had resumed office in 1973. However, Reuilf Steen, who was willing to open a door to the left, was elected to the chairmanship of the Labor Party in contravention of the traditional practice that the party chairman becomes prime minister when the party is in power.

This uneasy situation ended in January, 1981, when Nordli resigned and was replaced as prime minister by Gro Harlem Brundtland, who had been elected to the party chairmanship. She remained in office until October, 1981, when the Labor Party lost eleven seats in the Storting and a Conservative government was formed, with the support of a solid non-socialist majority of 80 seats. The left socialists raised their representation from two to four.

During her brief tenure in office, Brundtland did little to alter the direction of economic policy. She campaigned in 1981 on much the same platform as had her predecessors. Her defeat did nothing to allay the continuation of strife within the Labor Party on both domestic and foreign policy issues.[2]

The Labor Party had a number of economic policy goals that it pursued with vigor throughout the 1970s. Virtually all government policy documents during this period were prefaced by a restatement of these goals, which proved to be mutually inconsistent and had some unforeseen consequences.

Full Employment

"Work for all" was not just a general slogan for the Labor Party; it meant something much more specific:

> A generally tight labor market will not automatically assure that there

> will be appropriate employment for everyone who wants it. By using the means available under district policy, such as loans and arrangements or assistance, geographically differentiated taxation of employers to the National Insurance Scheme, and by using the law on the control of enterprise establishments and other means, there will be an attempt to create a more even distribution of employment opportunities.[3]

If at all possible, workers were not to be moved to places where jobs were available, but jobs were to be created where people were seeking employment. The 1980 National Budget stated flatly that "permitting a significant increase in registered unemployment, as had been the case in many other industrial countries, was not considered acceptable policy in Norway."[4]

The extent to which this goal was achieved emerges from the data in Table III-1. The depressions of 1974-75 and 1979-81 made only a small dent in the remarkable Norwegian performance, remarkable in the sense that most people would regard Norwegian unemployment as lower than necessary minimum frictional levels.

Getz Wold made the following comment on Norwegian unemployment rates (in 1981):

> . . . in one respect Norway is in a unique position. Our unemployment rate, at present less than 2 percent, is the second lowest in the world. The lowest registered rate is the Swiss. But I don't really accept that as comparable, because many of the Swiss unemployed have had to return to Italy as their labor contracts expired, and are not registered in Switzerland.[5]

The extent of Labor's commitment to full employment is all the more striking in light of the fact that a man or woman who became unemployed did not face a bleak future, but could count on a continuous source of income from a generous unemployment insurance scheme. Labor's concern was not that the unemployed would be thrown into poverty;

that is impossible in a welfare state. As the Ministry of Finance put it, "income from labor is an important basis for the standard of living, but work also has a value in itself for human development and emancipation."[6]

Table III-1. Unemployment in Norway and European OECD Countries, 1970-1984

(percent of labor force)

Year	Norway	European OECD Countries
1970	0.8	3.4
1971	0.8	3.5
1972	1.7	3.7
1973	1.5	3.5
1974	1.5	3.8
1975	2.3	4.9
1976	1.8	5.5
1977	1.5	5.6
1978	1.8	6.0
1979	1.9	6.2
1980	1.7	6.8
1981	2.0	8.5
1982	2.6	9.1
1983	3.3	10.2
1984	3.0	10.9

Source: OECD, Historical Statistics, 1960-1981; Main Economic Indicators, May, 1985

The policy has not been costless, as we shall see, but it persists. The Conservative government reaffirmed it, but with some qualification. Rolf Presthus, who became minister of finance in 1981, referred to "the broad consensus prevailing in Norway that one and all were entitled to have a job to go to, but this presupposed that the goods which were produced could be sold."[7] Unemployment was permitted to rise in 1983 and 1984 to levels unmatched in Norway in the 1970s, but still less than one-third as high as average rates in the European OECD countries.

Equality

As we have seen, at the beginning of the 1970s, Norway had one of the most equal distributions of income in the world. Yet one of Labor's major contemporary policy documents indicated that there was still a way to go:

> A continual equalization of actual living conditions and incomes will be an important task in the years ahead. . . . Income earners with especially low incomes should get a somewhat higher increase. Income-earners with a higher income than the average should also realize a certain increase, but the Government assumes that for groups which have a disposable income, for example, which is more than double that of manufacturing workers, the increase should be extremely limited.[8]

It has been remarked for Scandinavia as a whole that "equity is an almost tedious theme in social-democratic propaganda."[9] Its origins precede the socialist ascension to political power, but no party has been more zealous in pursuing it than the Norwegian Labor Party. The means employed to attain this objective were varied, including trade union collective bargaining, taxes, government transfer payments, and price control of basic consumer goods.

Expansion of the Welfare State

Labor's third major policy goal was the further perfection of an already comprehensive system of social services. There were warnings from some of the more cautious labor adherents. Getz Wold, who although not a member of the Labor Party, was close to it in social outlook, wrote:

> . . . experience has shown that other groups do not remain passive in the face of major shifts of income distribution in favor of certain groups. . . . When the political authorities want to bring about major changes in the distribution of income, it is imperative, first, to gain acceptance of the fact that when some groups are to receive a much greater than average increase in disposable income, other groups must

be content with correspondingly less.[10]

This and similar warnings went largely unheeded. Throughout the 1970s, social expenditures rose more rapidly than the national product, even though they had already constituted a large proportion of it at the beginning of the decade.

Monetary Policy

Subordinate to the three major policy goals just outlined but nevertheless of considerable economic importance was the maintenance of tight governmental control of investment resources, primarily through low rates of interest and the establishment of state-owned investment banks. Socialists appear to have a deep distrust of private bankers, and the Labor Party was no exception. There was a widespread belief that high interest rates provided the banks with too much power and profit. With low interest rates, credit had to be rationed, and the government could decide where it should go. This was also a potentially important instrument for redistributing income.

The OECD observed in 1975 that "it has been a salient feature of monetary policy in Norway that interest rates are kept low and are rarely changed. Moreover, the level as well as the structure of rates is regulated partly by the stipulation of certain rates . . . and partly as the result of an 'understanding' between the Authorities and the private banks, covering most lending and deposit rates."[11] The Bank of Norway put it more bluntly: ". . . the main purpose of the interest rate regulation was to ensure that certain borrower groups obtained loans on favorable terms for priority purpose."[12]

Labor succeeded in keeping interest rates low. For nine of the twelve years between 1970 and 1981, real long-term interest rates were negative in Norway, that is, nominal rates were below the rate of price inflation. Sweden, where low interest policy was also in vogue, could not match the Norwegian record. Even Switzerland, where large bank holdings of foreign deposits tended to depress interest rates, had higher average rates than Norway during the 1970s.[13]

It has become customary in Norway for the government to prepare an annual national budget for the coming year. This is a combination of an economic forecast and an indication of what the main outlines of government economic policy will be. The national budgets promulgated by Labor attest to its consistency of purpose. Thus, the 1972 budget promised continued development of educational and health institutions, a substantial increase in the grant budget of the state-owned Housing Bank, and a sharp increase in the capacity of the commercial banks to make loans for hospital and other social construction.[14] By contrast, the non-socialist coalition government that was in power briefly in 1972-1973 promised to "reduce the growth of government expenditures and to give adequate room for increased private demand. Such policy will slow the increase in the tax level and provide for a more rapid growth in the real disposable income of the working population."[15]

When Labor regained power, the earlier pattern was resumed. The 1976 budget proposed to institute a wide range of industrial subsidies to ensure full employment. The 1977 budget promised that "the central goal of government policy will continue to be to secure full employment and a reasonable income. In particular, the position of the low income groups will be improved."[16] In 1978: "The main objective of the Government's economic policy is to ensure full employment in spite of the fact that the outlook for the international economy is bleaker than originally assumed."[17] A year later, with international stagflation under way, the budget made the following commitment: "A main objective is to maintain full employment even though we may have greater problems on the labor market than at the present time. . . . Another major objective relates to an equitable distribution. The Government emphasizes that the tighter [anti-inflationary] measures shall not have an unreasonably adverse effect on weak groups."[18]

Neither the deepening of the recession nor the growth of inflation deflected Labor from the pursuit of its basic objectives. The 1980 Budget rejected unemployment as an anti-inflationary policy. Only on the eve of its electoral defeat in 1981 did Labor concede that "the different goals may be in conflict with each other and the fulfilment of various goals must be viewed in this light. . . . The desire to maintain employment at the same level in any one company or branch of industry in the short run may also easily come into conflict with the need for adjustments and profitable and secure jobs in the longer run."[19]

This realization came too late. In its single-minded

pursuit of its goals, the Labor Party inflicted considerable damage on the Norwegian economy, manufacturing in particular. Precisely how this came about is the theme of the rest of this book.

CHAPTER IV

WAGES AND EARNINGS

The Norwegian labor market is highly structured. Wages and other labor conditions are largely determined by a collective bargaining system of long standing. There is a high degree of job security, fortified by low unemployment. The government is itself a major employer, and has also played an increasingly large role in private sector bargaining.

Most blue collar workers, and many white collar workers as well, belong to one of the 34 national unions affiliated with the Norwegian Federation of Labor (LO). As of December 31, 1983, the Federation had 745,000 members, 38 percent of all employed persons. Approximately 350,000 were in independent associations, thus raising the percentage of organization to more than 55.

Employers are also highly organized. The Norwegian Employers' Confederation (NAF) claimed a membership of 10,208 enterprises, employing 389,500, in 1983. Employers were affiliated to NAF through 60 industry associations. An additional 165,000 persons were employed in enterprises belonging to independent employer associations.

The present bargaining system goes back to 1935, when the so-called Basic Agreement between the LO and the NAF was signed. As the system developed, there was a tendency for more authority to be vested in the central organizations. The central government bargains with the unions through the Ministry of Government Administration, while the Association of Local Authorities deals with unions on behalf of municipalities and other levels of local government.

In recent years, national agreements between LO and NAF have been negotiated every two years, and tend to set the pattern for the rest of the economy. On some occasions -- 1974 was one of them, 1982 another -- individual unions deal directly with their corresponding industry employer groups, leading to a wider spread of negotiated wage rates. The LO has considerable constitutional authority over its affiliated unions, but it has had to allow decentralized bargaining to relieve internal pressures.

Prior to formal negotiations, which normally take place in the spring, the parties begin discussions at two levels. A so-called Contact Committee, composed of the prime minister and several members of the cabinet, and representatives of various interest groups, including farmers and fishermen, provides a forum for examining broad policy issues. A Technical Expert Group, also with representatives of the interested parties, prepares brief reports on the economic situation to serve as a background for negotiations. It has been observed that "as an institution of consensus the [Technical Expert Group] appears to have been effective in providing an agreed economic base for collective negotiations and in eliminating the sort of suspicion of government economic forecasts that has surfaced in the course of incomes policy experiments in other countries."[1]

A practice that has caused increasing problems in recent years is the round of local bargaining that follows upon the national agreements. These are usually conducted at the level of the individual enterprise, and while they were originally intended to adjust national agreements to local conditions, they have taken on a life of their own. In some years, the average wage increases gained by the unions in local bargaining have exceeded national increases. Indeed, it has become customary to allow for what is termed "wage drift" in formulating national negotiation demands and offers.

The wage increases that have resulted from this dual bargaining system are shown in Table IV-1. Since 1977, wage drift has exceeded centrally negotiated increases in every year, and constituted two-thirds of total wage increases.

The size of the wage drift is attributable to a number of factors: the tightness of the labor market, the effect of national agreements on wage differentials, variation in the profitability of individual firms. National union leaders and officials of the employer associations dislike it, for it complicates the wage determination process. But bargaining at the enterprise level is popular both with the local union leadership and with enterprise management. The former are able to satisfy the wage demands of their members for a share in the returns to profitable firms, the latter welcome the opportunity to attract and hold labor, particularly skilled labor, in tight labor markets. Wage drift is a correction to wage determination that does not conform to market reality. The greater the disequilibrium created by national bargaining, the greater the wage drift adjustment.

Table IV-1. Annual Wage Increases for Adult Males, by Collective Agreement and Wage Drift, 1970-1984

(percent increase, first quarter to first quarter)

Year	Hourly Wage Increase	Increase -- Due to Collective Agreement	Increase -- Due to Wage Drift
1970	6.0	0	6.0
1971	15.6	8.7	6.9
1972	9.3	4.4	4.9
1973	9.6	3.8	5.8
1974	12.2	4.9	7.3
1975	25.2	17.2	8.0
1976	13.0	8.9	4.1
1977	15.6	9.4	6.2
1978	10.0	2.9	7.1
1979	6.0	2.1	3.9
1980	2.7	0	2.7
1981	13.2	5.0	8.2
1982	8.5	1.7	6.8
1983	11.4	5.4	6.0
1984	6.6	0.9	5.7

Sources: 1970-1977: Central Statistical Bureau, Economic Survey, 1982, p. 61
1978-1984: Revised National Budget, 1985, p. 13

The Skanland Report

Norway began to experience mounting inflation in the early 1970s. Consumer prices were rising about 7 percent per annum, and wages even more rapidly. The relevant data appear in Table IV-2. The government appointed a committee to look into the situation. It was chaired by Hermod Skanland, deputy governor of the Bank of Norway, and included representatives of the major interest groups and political parties.

The committee report drew attention to a number of factors that were contributing to price inflation, including excess demand, a rise in world market prices, larger government income transfers, and wage increases in excess of productivity. It set out some basic principles necessary for a successful policy of limiting incomes:

Table IV-2. Nominal and Real Hourly Earnings in Manufacturing, Norway, 1970-1983

(year to year percentage changes)

Year	Nominal[a] Earnings	Consumer Prices	Real Earnings
1970	12.0	10.6	1.3
1971	12.5	6.2	5.9
1972	9.1	7.2	1.8
1973	10.6	7.5	2.9
1974	17.5	9.4	7.4
1975	20.2	11.7	7.6
1976	17.0	9.1	7.2
1977	10.9	9.1	1.7
1978	8.0	8.1	-0.1
1979	2.9	4.8	-1.8
1980	9.8	10.9	-0.9
1981	10.4	13.6	-2.9
1982	10.9	11.4	-0.4
1983	8.2	7.9	0.3
1984	11.1	6.6	4.5

Source: OECD, Main Economic Indicators

[a]These data are not directly comparable with those in Table IV-1 due to differences in employee groups covered and the time periods.

1. Any incomes policy must encompass a wide range of incomes, not just wages and salaries.

2. Although basic policies would ultimately be made by the political authorities, there should be extensive cooperation with organizations sharing responsibility for implementation.

3. The targets must be realistic.

4. While the central purpose would be price control, emphasis should also be placed on income distribution effects.

5. Any policy would have to be long-term in character.

6. Other aspects of economic policy would have to be directed toward the same objectives.

The report proposed the establishment of a formal Council for Prices and Incomes which would include governmental and interest group representatives. It argued that "the traditional view that wage negotiations [between LO and NAF] are a question of distributing real resources between two opposing parties is actually not true." Private decisions could be frustrated by taxes and government transfer payments. A prerequisite for price stability was that targets set by government and by the labor market organizations "are mutually compatible and that agreement can be reached on the harmonious use of the measures at the disposal of each group."

The new Council would establish a number of targets, including purchases of goods and services by the public sector; the relationship between wage and entrepreneurial income; disposable real income for wage earners; price trends; and expectations for income other than wages. In essence, there would be negotiated guidelines. While they would have to be submitted to the Storting for approval, it was hoped that informal consultation would minimize the danger of rejection by that body.

This ambitious blueprint would have enhanced the authority of LO and NAF and reduced the possibility of decentralized bargaining. Although the LO representative on the Skanland committee signed the report, the LO yielded to pressure from its constituent unions and refused to approve it, thus effectively killing it. The Labor government, which favored the recommendation, was not prepared to override its trade union allies.

What looked like a promising solution to some vexing institutional problems was thus vetoed by the national unions. Both the Labor Party and LO were later to regret this missed opportunity. Skanland himself summarized the alternative that was chosen by default:

> I believe that if anyone living under a system different from the one actually existing in our country had suggested that prices and incomes should be determined by interest groups enjoying a more or less monopolistic position in the market, while the authorities should be responsible for maintaining full employment, the proposal would have been rejected as inconsistent, inflationary, and completely devoid

> of any promise of optimal use of productive resources. Nevertheless, such is the system under which we live.[2]

The 1974 Wage Explosion

The year 1974 witnessed the first episode of decentralized collective bargaining since 1961. Every national union was permitted to bargain on its own behalf. The government was committed to maintain full employment, and indeed, the rate of unemployment in 1974 was only 1.5 percent. The euphoria of the oil discoveries had begun to spread. Real wages had been rising; the average increase for the four preceding years had been almost 3 percent. But the left had scored important electoral victories in 1973, and both the Labor Party and the trade unions were being pressed by the demands of their constituents for higher incomes, which the more responsible leadership found it hard to resist.

The metal and engineering industries, which were particularly short of labor, set the pattern by negotiating a 6.3 percent wage increase as of April, 1974, plus an additional 5 percent the following year. There were additional increments averaging 5 percent for low wage industries. And as a step toward further wage equality, the metal and engineering trades instituted a wage guarantee whereby no worker would earn less than 87 percent of the national average in his craft. Index adjustment for future price increases was also provided.

On top of all this, wage drift reached the highest level ever recorded in Norway. This was due not only to the tight labor market, but also to the egalitarian profile of the settlement. For example, low paid workers in the canning industry received a 17 percent increase in the first contract year compared with the 6.3 percent increase for the metal workers. The 87 percent guarantee within the metal trades also upset traditional wage differentials. The results were predictable; skilled workers simply convinced their employers, many of whom were anticipating profitable operations and did not need much convincing, to pay them in excess of what the national agreement had stipulated. Manufacturing earnings rose by 17.5 percent in 1974 and 20.2 percent in 1975, sufficiently in excess of prices to provide very substantial real wage increases (see Table IV-2).

In reviewing the 1974 settlement, the OECD concluded

that it was "clearly outside the range experienced in the past, possibly pointing to certain structural changes or an upward shift of the Phillips curve."[3] In retrospect, however, there was nothing as complicated as a change in the unemployment-wage relationship. It was simply that Labor's commitment to full employment and income equality, plus a labor movement beset with internal political problems, had created ideal conditions for economically unsound demands that employers were powerless to resist, even if they had been so minded.

Although there was no national bargaining in 1975, the 1974 contractual provisions, cost of living indexing, and more wage drift, resulted in another large wage increase. Even at that, the full force of the wage inflation does not show up in the data, for in lieu of a portion of the amounts that were payable because of indexing, the government reduced withholding taxes, raised children's allowances, and increased food subsidies. Prices were frozen from September 1, 1975, to the end of the year, and profits were squeezed by forbidding the employers' share of the index adjustment to be reflected in price increases.

The 1974 wage explosion was a major factor behind the inflationary forces during the rest of the decade. It is true that the Norwegian experience was not unique; manufacturing wages also rose sharply in Denmark, Sweden, Britain, and Germany. However, real wages increased considerably less in those countries than in Norway; for them, it was a question of wages chasing prices, while in Norway, wages led the way.

Coordinated Wage Settlements

With the 1974 experience in mind, the government initiated the 1976 negotiating round by inviting the labor market organizations and farm representatives to help it fashion a settlement that would dampen inflation. Alternative models of wage, price, and tax changes were prepared to enable the participants better to appreciate the consequences of their demands and to bring more consistency into the eventual outcome. Prior to the expiration of the old contracts, the government presented a proposal which provided a basis for the final negotiations. The eventual agreement was accepted by referenda among union members and by employers, and by the Storting insofar as there were budgetary implications.

On the wage side, there was to be full compensation

for a reduction in working hours from 42½ to 40 a week, plus an average increase of 4 percent. Farmers received a substantially larger income increase. Cost of living indexing was dropped, though a possible contract reopening was provided for after the first year of the two-year agreement. For its part, the government agreed to reduce income tax rates; to lower the employers' social security contributions; to raise family allowances by 8 percent, and social security payments by from 5 to 7 percent. It also agreed to keep professional income and profits within the wage guidelines. An egalitarian profile was provided by proportionately greater tax reductions for taxpayers in the lower income brackets.

Most OECD countries attempted to curb inflation by various deflationary measures, including reduction of government expenditures and permitting unemployment to rise. But the Norwegian government was able to adopt this comprehensive and expensive incomes policy approach because of the oil sector. Trade deficits in 1975 and 1976 did not act as constraints because of the knowledge that the oil would soon start flowing.

Ordinarily, a policy that resulted in a 3 percent increase in real disposable wages, tax reductions equivalent to 2 percent of the GDP, and sharply increased payments to farmers and pensioners would not recommend itself as an ideal way to fight inflation. The traditional solution would have been to restrain demand by raising taxes and hoping that a consequently looser labor market would dampen wage drift. However, an election was coming up in 1977, and expanding demand had many political advantages. Unemployment actually declined from 1976 to 1977, and the unions appeared to be satisfied with the settlement, as were the farmers. The problem was that price inflation continued.

The year 1977 was an off-contract year, but the government was nevertheless obliged to step in once again. The unions had the right to reopen the contract, and they did so, demanding a nominal wage increase of 5.6 percent. They were induced to accept 2.9 percent against a government commitment to reduce taxes further, yielding a 2.5 percent increase in real disposable income. However, wage drift went up sharply, so that in total, there was a substantial increase in nominal wages, which is what the government had hoped to avoid.

A Change of Course

With an electoral victory behind it, the Labor Party began to employ more orthodox weapons in its fight against inflation at the end of 1977 and during 1978. Interest rates were raised and credit availability reduced; the krone was devalued in order to stimulate exports; and price control instituted. It was finally realized that the country could not spend its way out of inflation.

The government did not immediately become involved in the 1978 contract negotiations. The LO asked for a minimum wage guarantee for low paid workers and the maintenance of real wages for all. The NAF did not see any basis for higher wages, and came out against any increases or guarantees. When the LO persisted in its demands and threatened to call a general strike, the Storting enacted legislation providing for compulsory arbitration, perhaps to save the LO leadership the embarrassment of compromising in the face of rank and file demands that had been stimulated by the double-digit inflation of 1977. The arbitration board limited the wage increase to $1\frac{3}{4}$ percent plus an additional increase for low paid workers.

Despite this modest award, industrial earnings continued to rise sharply in 1978, impelled primarily by wage drift. The basic problem was that the government was not prepared to take the ultimate step of letting the labor market move toward greater slack. As the OECD put it, "Advance announcements that the Government would finance a large share of the guaranteed increases in real disposable wage income may have been partly responsible for the fact that actual real income gains turned out to be stronger than those thought compatible with healthy overall economic equilibrium."[4]

With unemployment averaging only 1.8 percent in 1978, Finance Minister Kleppe announced a special employment program to assist those few unemployed who had fallen through the very fine safety net. Vocational training opportunities were increased, as were sheltered employment positions. A scheme under which anyone over the age of 50 years who had been unemployed for some time could have one-third of his wages subsidized for up to six months was introduced. There were new youth employment grants and a pilot scheme of "equality" grants to improve the labor market position of women. All this was on top of a very "active" labor market policy.[5]

These measures, together with a number of other factors, contributed to the maintenance of an extremely tight labor market. Firms were reporting that labor shortages prevented output growth. There had been an extraordinary growth in the demand for construction workers because of the oil industry. There was a geographical imbalance in labor supply and demand. The flow of labor out of agriculture was reduced because of large subsidized income increases in that sector. And the government was also subsidizing jobs in unprofitable enterprises.[6]

The fundamental problem was the lack of labor mobility reflected in the implausibly low unemployment rates. The director of the employment service pointed out that "geographical mobility has gradually diminished over time, particularly the last 15-20 years. It is also less acceptable politically. It is more difficult when both husband and wife have jobs. More than in the past, people own their own homes, which they would have to leave if they moved."[7] Political unacceptability was the key. In pursuit of absolutely full employment, without the obligation to move, the Labor government used government funds to move jobs to pockets of unemployment. This policy had few parallels; even Sweden, famous for its active labor market policy, stimulated mobility through travel grants and settling-in allowances rather than subsidizing unprofitable positions.

The year 1979 was the quietest on the Norwegian labor market during the entire decade of the 1970s. There were no contract negotiations, and wage drift fell to its lowest level in many years. This was brought about by a freeze of prices and incomes initiated in September, 1978, and lasting to the end of 1979. Other measures having proven ineffective, Labor used the ultimate means of intervention in both labor and product markets.

Coordinated Wage Settlements Again

The predictable problem for 1980 was to prevent pent up inflationary forces from undoing the price and wage stability that had been achieved by the freeze in 1979. Labor would not consider any slackening of the labor market. Unemployment had been maintained at 1.9 percent in 1979, whereas it had risen to 6.2 percent in the European OECD countries, and to 2.1 percent even in Sweden. Resumption of coordinated wage settlements seemed to be the best course to follow.

The parties adopted a proposal by the state mediator calling for a wage increase target of 8 percent for the 1980 negotiated round, to be accomplished by a contract increase of 5 percent and a 3 percent wage drift. It was agreed that if this figure were exceeded, a wage drift ceiling would be imposed for the second year of the contract. For its part, the government committed itself to a one percent reduction in the income tax, an increase in children's allowances, and the introduction of tax-deductibility of trade union dues.

The most interesting feature of the 1980 agreement, however, was a unique experiment in wage equality. No firm within the LO-NAF bargaining area in manufacturing would be permitted to pay average wages below 85 percent of the average wage for the entire bargaining area. Half of any consequent wage increase was to be subsidized out of a special low-wage fund financed by the employees. This scheme was intended to give Norway a wage structure unmatched anywhere in terms of compression of differentials.

Wage-price stability proved to be elusive. One of the problems was rising expectations resulting from the second OPEC oil price increase. A resumption of negotiations in the spring of 1981 was mandated by a contract reopening clause in the 1980 agreement.

There had been a change in the government financial system involving a reduction in income taxes and consumer subsidies, while indirect taxes, including excises on cars, tobacco, and alcohol, were raised. It was understood that through collective bargaining, the regressive impact of the change would be corrected. However, LO refused to agree to a ceiling on wage drift, which benefited mainly the better paid worker, despite the 1980 agreement. Direct negotiations broke down, and compulsory arbitration was imposed once again.

The arbitration board granted an average wage increase of 9.65 percent, and imposed a ceiling on wage drift, with some exceptions for enterprises that gave regular productivity bonuses. LO appealed this decision to the Labor Court, which is the court of final appeal for all contract grievances, but the appeal was rejected. This did serve to reduce wage drift, although it remained at a fairly substantial level, amounting to 6.8 percent from the first quarter of 1981 to the first quarter of 1982.

This settlement was the last that took place during

the tenure of the Labor government. Under the Conservatives, the 1982 negotiations reverted to decentralized bargaining with a minimal government role. Agreement was reached on an average wage increase of 4.5 percent, plus one percent to allow for adjustments in seniority, overtime, and night shifts. The wage drift ceiling was partially removed.

The low wage guarantee was retained for those firms already subject to it, but with a significant change. Instead of selecting every enterprise that could not meet the 85 percent minimum, the determination was made on the basis of the industry branch to which the enterprise belonged. Only if the average wage paid in that industry was below 85 percent of the average for all industry would an enterprise be obliged to grant the guaranteed wage. For example, if within the canning industry, half the firms were paying 90 percent of the general average, and the other half 80 percent, there would be no extra pay. LO agreed to this change largely because many of its members objected to paying into the low wage fund. Any new wage increases were to be paid entirely by the employers, with no further subsidies from the fund.

Results of Labor's Wage Policies

The data in Table IV-3 provide some indication of what happened to inter-industrial wage differentials during the 1970s. Two of the five lowest wage industries in 1971 had improved their relative positions by 1982, while two had retrogressed. At the other end, only one industry, printing and publishing, had drawn further ahead of the general average, while three had moved closer to it. If one took an unweighted average of change in all industries, the overwhelming impression is one of stability in the industrial wage structure.

This kind of comparison has its pitfalls. Some industries grew during the decade, while others, such as shipbuilding, declined. However, the impression that the drive for equality of wages was not very successful in this respect is reinforced by looking at broad industry groups. Average wages for the metal trades as a whole were 4 percent above the general industrial average in 1971, and 2 percent above in 1982. Textiles and clothing changed but little, while the average for pulp and paper rose only from 98 to 99 percent of the average.

Table IV-3. Industrial Wage Differentials, Norway, 1971 and 1982

(percent of average industrial wage)

	1971	1982
Five lowest wage industries in 1971		
Sawmills and planing mills	85	89
Clothing	85	83
Textiles	86	86
Shoes	87	78
Dairies	87	92
Five highest wage industries in 1971		
Shipbuilding	107	103
Foundries	109	97
Cement and concrete	111	101
Printing and publishing	113	130
Aircraft repair	119	119

Source: Central Bureau of Statistics, *Statistical Yearbooks*

Much of the same picture emerges from data on skill differentials (Table IV-4). Only in the case of hotels and restaurants was there a narrowing of differentials, which were almost unchanged for the engineering and construction industries. It might be argued that in the latter, the skill differentials were already so narrow in 1971 that little change could be expected. But this did not keep the labor movement from trying.

Only in the case of male-female wage differentials do the data suggest significant compression (Table IV-5). Part of the change may have been due to the movement of women into male-dominated, high wage industries, but there is no question that Labor's drive for income equality benefited women.

An explanation of why Labor failed to make much progress in equalizing differentials was provided in a speech by Tor Halvorsen, who was chairman of the LO during much of the period under consideration. It is worth quoting at some length:

Table IV-4. <u>Wage Differentials by Skill, Norway, 1971 and 1982</u>

(unskilled as a percentage of skilled wage)

Industry	1971	1982
Engineering, adult males	91.5	92.5
Construction		
Oslo	85.4	85.0
Rest of country	86.1	85.4
Hotels and restaurants (cooks versus kitchen and pantry staff, men, Oslo)	76.3	81.0

Source: Central Bureau of Statistics, <u>Wage Statistics</u>

Table IV-5. <u>Wage Differentials by Sex, Norway, 1971 and 1982</u>

(female as a percentage of male earnings)

	1971	1982
Manufacturing	75.4	83.2
Private land transportation	80.5	87.1
Hotels and restaurants	84.9	88.7
Wholesale and retail trade	65.9	76.4
Office employees[a]	60.4	65.3

Source: Central Bureau of Statistics, <u>Wage Statistics</u>

[a]Establishments affiliated with the Norwegian Employers' Federation

It is the enterprise itself that must bear the wage costs of its employees. . . . We must eventually plan to wind up the low wage fund, and regard it as a transitory phenomenon that we have used to raise the low wage group to a reasonable level.

This point of view has an organizational side that cannot be neglected in the real world. The tax that wage earners covered by the LO-NAF contract pay . . . is a burden for us, particularly in areas where there are competing organizations that do not have such obligations. . . .

We must take several things into consideration with respect to the guarantee percentage. The guarantee must secure a minimal level for the people. On the other hand, there are obvious limits as to how far we can stretch the guarantee. Differentials that are too narrow result in a poor economic structure in the long run. 85 percent is a reasonable level. If we want greater compression, it should be done by means other than wage guarantees.

Another side of the low wage question is the relationship between high and low wage employees. In the trade unions, this dilemma has often been presented to us during local wage negotiations, where our members in strong enterprises point out that it does not help low wage workers if they refrain from taking wage supplements. Then the enterprise simply has a greater income. . . .

There are some relationships to which we must pay attention. First, wage equalization provides low wage workers with higher wages. Seen in isolation, that is good. But we cannot neglect the fact that it creates problems for the unskilled, for young workers, and for others who are not attractive employees. As long as they have lower wages than

> other workers, they can find employment, but if their labor becomes considerably more expensive, it will be less tempting for employers to hire them.
>
> Another dilemma is that our economy and our working life simply require some wage differentials. It is important for the economy that labor is employed in such a way as to make the best possible use of it. . . . There is therefore a contradiction between the egalitarian wage policy based on our fundamental ideal of equal living standards and the signals that the market gives to guide labor where it can best be used. For this reason it is important that the egalitarian wage policy be supplemented by other instruments, for example, tax policy and regulations respecting labor mobility which do not contradict market signals to the same extent.
>
> The problem can be put this way: many work in enterprises with low ability to pay wages. It is difficult -- if not impossible -- to solve the low wage problem unless the enterprise has sufficient capacity to pay. . . . We will fight to protect the low wage guarantee, but at the same time we must be willing to see that such a policy has its price.[8]

It would be difficult to find a clearer analysis by a trade union leader of the pitfalls of artificially-compressed wage differentials. Wage drift was bound to frustrate the drive for egalitarianism in a situation in which collective agreements do not stipulate the wages that are actually to be paid, but merely establish a minimum that can be adjusted upward by local negotiation, sometimes between the employer and the individual employee. This system does have the advantage of permitting the restoration of some semblance of market wage determination and thus preventing the misallocation of labor that is bound to result from unrealistic wage differentials. On the other hand, it does make the calculations necessary to the formulation of an incomes policy very difficult. It may be that the

Norwegians are becoming adept at estimating the wage drift correction factor, and allowing for it in advance. But this seems like an unnecessarily complicated system of wage setting.

Outside the LO-NAF Bargaining Area

While LO bargains with NAF on behalf of most blue collar workers and many white collar workers as well, most of the latter group are represented by organizations outside the LO, or are independent professionals whose income is affected by government regulation or by the biennial LO-NAF settlements. LO efforts to secure a greater proportion of the national income for its members on the ground that they are the least affluent groups in society are not always appreciated on the outside.

To a considerable extent, non-LO organizations represent employees in economic sectors not subject to direct foreign competition. This should give them more bargaining power, but in fact, they did not do as well as LO members during the 1970s.

A number of things led to this result. In 1980, for example, the Joint Academic Association, composed of professions requiring an academic degree, from architects to tax officials, struck against a coordinated settlement, and the large teachers' union threatened to strike. Both were blocked by a compulsory arbitration decree, although the LO and NAF were permitted to consummate an agreement without intervention.

The pattern of tax adjustments that the government contributed to the coordinated settlements tended to favor the lower paid at the expense of those with higher incomes. For example, the Kr. 90,000 after tax income received by an industrial worker with two children in 1980 represented a real increase of 20 percent over 1970. A works manager with the same family circumstances earning Kr. 125,000 in 1980 would have had no increase in real disposable income over the decade, while a senior engineer earning Kr. 175,000 would have suffered a 12 percent loss.

The LO has taken the position that it is inequitable to permit outside groups to strike for their demands. The argument is an interesting one:

> In today's society, and with the current system of collective bargaining, the right to strike is

> in many ways a realistic right only for those groups which negotiate outside the main bargaining area. Formally we have equal rights, but in reality our social responsibility makes greater demands on large groups than on small special interest groups in this regard.[9]

Collective Bargaining in Retrospect

Norway had an excellent record of industrial peace during the 1970s. A powerful trade union movement closely allied with a socialist government has no need to engage in strikes to achieve its objectives. The LO leadership had to assume some responsibility for the conduct of the economy because there was no real alternative. If it gave full rein to its constituent organizations, as it did in 1974, the consequences could be severe.

Top union leaders are not always free to make the decisions that they regard as consistent with the long run interests of their members. They must stand for reelection in the short run, and they are subject to conflicting current pressures. The increasing availability of oil revenues in the 1970s did not make their task easier.

Under such circumstances, responsibility devolved upon the government to reconcile labor market conditions with the requirements of sound economic development, either by intervention or non-intervention. The Labor government did neither. Ideology triumphed over pragmatism. It stifled the mobility that is essential to a well functioning labor market. Real bargaining was handicapped by removal of the threat of economic consequences for both sides of the bargaining table. Workers were assured that their jobs were secure no matter what their demands. Employers could count on subsidies if they agreed to uneconomic conditions of labor. The difficulties were compounded by the quest for greater equality in an already equal society.

In 1971, the OECD, in reviewing the institutional framework of the Norwegian bargaining system, concluded that it worked reasonably well. Wage drift was large but fairly stable -- 3 to 4 percent a year. The government intervened only on rare occasions, the Contact and Expert committees helped provide the bargaining parties and the general public with objective information on the economic framework.[10]

More recent reviews have been less complimentary. On equality: "Since personal income distribution is more equitable in Norway than in most other OECD countries, a slowdown in redistributional efforts through wage policy may contribute to a lower wage drift."[11] And on the system in general: "It would therefore seem that the income formation process is no longer well suited to the new environment of lower growth and increased rigidities in which the Norwegian economy has to operate."[12]

A number of things contributed to the deterioration of the system, not least the oil discoveries. But politics cannot be neglected. Prior to the 1970s, the Labor Party, though divided on foreign policy, was reasonably united on domestic policy issues. It had a vigorous leadership and dealt with the LO from a position of strength. As its parliamentary power waned, internal fissions began to develop. There was a growing challenge from the left, culminating in the Common Market referendum. The Labor prime minister had announced that he would resign if the vote for membership were negative, and he was backed by the LO. Nevertheless, prominent Labor politicians campaigned against the Common Market, and helped bring down their own government.

The more conservative Labor Party leadership seemed to lose the will to resist the pressure from the left. Utopian slogans replaced the pragmatism of the past. The trade union leadership could not rely upon the government to rescue it from importunate membership demands, even if it had wanted such assistance. On the contrary, the government added fuel to the flames by financing unwise wage settlements.

It may be that the non-socialist coalition that came to power in 1981 will bring about some changes if it remains in office long enough. It did not intervene in the 1984 biennial negotiations, which were again conducted on a decentralized basis. Wage increases averaged over 6 percent for the year, mainly as a result of wage drift. The low wage guarantee was extended for two years, but it was agreed that LO and NAF would set up a joint committee to examine the entire arrangement. In return, the two organizations were to consider the economic feasibility of reducing working hours to 37½ a week, in time for the 1986 negotiations.

Despite the absence of LO and NAF from the wage

negotiations, there was no wage explosion, and the fact that individual unions made separate bargains provided a more favorable framework for curbing wage drift. It remains to be seen whether this pattern marks a permanent shift in the system.

CHAPTER V

EMPLOYMENT AND UNEMPLOYMENT

The combination of oil discoveries and economic policies produced some changes in the structure of employment. There should have been more. The incentives for mobility were perverse; they discouraged rather than encouraged the movement of employees from one part of the country to another, from one industry to another, from one employer to another.

Total employment increased from 1,694,000 persons in 1972 to 1,970,000 in 1984, an increase of 276,000. To a considerable extent, this increase was due to a rise in the propensity of women to work. In 1972, women constituted 36 percent of employed persons; in 1984, 43 percent. Looking at it another way, the participation rate for men (the ratio of men in the labor force to the male population of working age) fell from 88 percent to 75 percent over the period, while that for women rose from 49 percent to 57 percent.

As in other advanced countries, Norwegian employment moved toward the services and away from traditional industries. Between 1972 and 1984, the manufacturing sector lost 29,000 jobs and fell from 24 percent to 18 percent of total employment. Community, social, and personal services were the big gainers, with 258,000 additional jobs and an increase in the percentage of total employment from 24 to 33. Most of this increase was actually in the public sector, which accounted for about two-thirds of the total increase in employment since the mid-1960s. Within the public sector, health and educational employment grew most rapidly. These two sectors alone employed 146,000 more persons in 1984 than in 1972. They were the most significant employment growth centers of the welfare state.

Direct oil activities do not provide much employment. In 1983, this sector provided only 7,100 man-years of employment, about ½ percent of total employment. However, if such oil related activities as refineries, construction of platforms, and base supplies are included, there were 52,000 oil jobs, more than three percent of total employment. Many of these jobs tended to be higher skilled and higher paid, raising labor market tensions.[1]

The principal employment gainer within the

manufacturing sector was machinery and equipment, much of it related to the oil industry. The decline in exports of such traditional industries as wood processing, chemicals, rubber, plastics, and mineral products contributed to employment stagnation in those industries.

Unemployment was maintained at very low levels (see Table III-1 above) by a combination of an expansionary demand policy and a number of special programs, designed to eliminate any vestiges of structural unemployment. In 1984, some 34,000 persons were engaged in these programs (1.7 percent of the labor force), which included temporary government employment, training, youth subsidies, and grants for the long-time unemployed over 50 years old. It might be noted that the number of people who benefited from these programs more than doubled in the first three years of Conservative government (1981-84), in response to the rise in unemployment -- and to the forthcoming general elections.

Norwegian labor markets tend to be rigid and segmented. A study of the structural problems of industry took special note of low labor mobility:

> Norway has a special problem which is perhaps greater than in other countries because the labor market is so divided into small geographical units. There are many reasons why mobility is not greater. There is a desire to maintain the residence pattern and people feel themselves tied to their homesteads and place of work. It is often not easy to find a new home. In addition, social security regulations are such that people will seldom feel obliged to move.[2]

There is strong resistance in Norway to the idea that anyone should be obliged to move in order to secure employment. Various arguments are advanced in support of this view: the country is large, so that migration means the cutting of social ties; construction costs are high, so that mobility may entail a loss of housing and public building investment; undue concentration of industry and population may create social and economic problems; it would be undesirable to depopulate remote areas for national defense reasons.

In fact, almost three-quarters of the Norwegian population is located in the southern part of the country, with good means of communication. The distances involved in moving within this area would not be regarded as burdensome in the United States or in a good part of Europe. Housing is an obstacle to movement wherever home ownership is common. Norwegian housing costs cannot be so much greater than comparable costs in the major cities of industrialized countries. Personal hardship involved in moving is likely to be less in Norway than in many other nations because of the homogeneity there of language, religion, and social customs. And if national defense is at stake, the cost of immobility should be budgeted to military, not employment, expenditures.

In an interview in 1982, Petter Jakob Bjerve, a former Labor minister of finance and director of the Central Bureau of Statistics, set forth an excellent summary of the mobility problem:

> . . . changes in industrial technique and in the composition of consumption make it impossible to prevent adjustments which will require the mobility of manpower both geographically and among enterprises. Geographical adjustment has met the greater resistance. But without both types of movement we will have lower living standards and higher unemployment in many parts of the country than we could have if we were willing to move. I must say that politicians place too much stress on the idea that one can always live where he is born.
>
> Some years ago, the Central Bureau of Statistics undertook a study in which people who moved were asked what they thought of it. One of the most interesting results was that most were not sorry that they had moved. We cannot have both growth and frozen residence patterns, even if there are other grounds to maintain dispersed employment, such as national security and local policy considerations.[3]

Public Employment

Reference has already been made to the expansion of service employment. Public employment in particular experienced very rapid growth. From 1972 to 1984, it rose from 18 to 27 percent of total employment. These figures do not include employees in postal and telecommunications services, in publicly owned railroad and bus transport, in the state financial institutions, in electricity, gas, and water supply, and in government-owned enterprises such as Statoil and Norsk Hydro. If these were added, government employment might rise to one-third of total employment.

The welfare state is a voracious consumer of employment. Its expansion requires a growing bureaucracy. Productivity in services tends to rise less rapidly than in the production of goods, and the government sector is likely to lag behind even the private service sector. The implication is that the expansion of the welfare state and the concomitant growth of government employment is a factor tending to lower the rate of economic growth.

Toward a More Flexible Labor Market

The non-socialist government has proposed some measures designed to raise labor mobility, apart from permitting unemployment to rise. For example, legislation had been enacted in 1971 barring the operation of firms that supplied temporary workers. This has been changed to permit specialized employment firms to function, but only for commercial and office work.

It was also recognized that liberal administration of the unemployment insurance laws permits employers, through reduced working hours and layoffs, to effectuate temporary curtailment of their staffs at public expense. However, the government decided that there was little it could do to remedy the situation because of lack of adequate staffing of the state employment service.

Other areas in which Norway is attempting to loosen administrative regulations that tend to reduce the availability of labor involve overtime work, night work, and work on Sundays and holidays, all of them strictly limited.[4]

Norway appears to have a relatively immobile labor force. Lack of comparative statistical data makes it difficult to document this statement, but few countries impose more obstacles to labor mobility and have adopted so many measures that encourge immobility. It should be emphasized that the various laws and regulations in effect were intended to promote the welfare of the individual employee, with the social cost an unintended by-product. Enabling individuals permanently to remain at particular places of work does provide benefits, but these may be outweighed by the costs entailed by a less dynamic economy and a consequent loss of goods and services. The Norwegian tradeoff is diametrically opposed to practice in the United States, for example, where people are accustomed to move fairly frequently, often for long distances, in pursuit of better-paying and presumably more productive jobs. Norwegians may prefer to avoid change, but they may not have been able to afford immobility were it not for the fortuitous discovery of oil in the North Sea.

CHAPTER VI

MANAGING THE ECONOMY

Norway achieved a relatively favorable rate of economic growth throughout most of the period with which we are concerned. The average annual rate of GDP growth from 1972 to 1982 was 3.9 percent, compared with 2.2 percent for the European members of the OECD. However, without the oil activities (and shipping, which reduced the growth rate for some years) the growth for the period was 3.0 percent.[1]

When the first oil price increase in 1974 sent much of the West into recession, Norwegian industry continued to operate at a high level of capacity utilization. There was steady growth from 1975 to 1980, but the international depression that began in 1980 hit Norway with full force and brought economic growth to a halt in 1981. The economy picked up in 1983 and 1984 with GDP growth rates of about the same magnitude as the average for the 1970s.

It is difficult to be precise about the contribution of oil to the growth of the Norwegian economy since oil output is only part of the story. Oil sector investment, among other things, triggered a boom in manufacturing industries related to oil and had a multiplier effect. The figures quoted above should be regarded as approximations.

Fiscal Policy

To say that Norway has long been a high tax country is to say nothing that is not well known. The interesting question is why it remained near the top of the international tax league when oil revenues began to fill government coffers.

The data relating to taxes as a proportion of the GDP are shown in Table VI-1. Norway entered the exclusive 50 percent club, whose only members are the other Scandinavian countries and The Netherlands, in 1980, although it has retreated since then under the Conservative government. The share of transfer payments did not change greatly over the decade.

In international perspective, Norway has had relatively high taxes on consumption, private income, and corporate profits, and low taxes on property. Taxes on expenditures have averaged 17 percent of GDP compared with an OECD average of 10 percent, which has contributed to inflation.[2]

Table VI-1. Gross and Net Taxes as a Percentage of the Gross Domestic Product, 1971-1984

Year	Gross Taxes	Transfers to the Private Sector	Net Taxes	Gross Taxes without Oil Taxes
1971	44.7	20.0	24.6	48.9
1972	46.3	21.0	25.3	50.4
1973	47.4	21.3	26.1	52.0
1974	45.6	20.7	24.7	50.0
1975	46.6	21.8	25.0	50.0
1976	47.6	22.5	24.7	50.3
1977	47.8	20.9	26.9	50.5
1978	48.1	22.2	25.9	51.0
1979	47.8	22.0	25.8	50.7
1980	50.2	20.9	29.4	52.4
1981	48.7	21.0	27.7	51.5
1982	48.5	21.3	26.9	50.7
1983[a]	47.9	21.6	26.5	49.8
1984[a]	47.3	21.7	23.6	49.3

Source: National Budget, 1978, p. 99; Revised National Budget, 1984, p. 86; Revised National Budget, 1985, p. 97

[a]Preliminary

As for marginal income tax rates, Norway has been second only to Sweden. Their evolution during the 1970s is illustrated in Table VI-2. There was a decrease for taxpayers in the lower brackets, where most wage earners would have been found, but a substantial rise for persons at the top of the income distribution. Norway is one of the few countries in which interest payments are fully deductible from taxable gross income, thus reducing somewhat the tax burden borne by higher income groups. It also provides a strong incentive for private borrowing.

It was Labor's policy, when tax brackets were adjusted downward as part of coordinated wage settlements, to favor lower income groups. However, they did little more than keep up with the bracket creep induced by inflation. As early as 1974 it was recognized that tax reform was needed; Per Kleppe, the minister of finance, stated that "it was obvious that when complaints about high marginal tax rates had spread even to ordinary wage earners, the time for reform was ripe."[3] Some relief was given to lower income groups, but on the other hand, the so-called 80 percent rule, which ensured that taxes could not exceed 80 percent of total income, was repealed.

Table VI-2. Average and Marginal Income Tax Rates, Norway, 1973 and 1982[a]

Gross Wage Income, 1981 Kroner	Average Tax Rates		Marginal Tax Rates	
	1973	1982	1973	1982
Single person				
50,000	28.1	27.2	39.7	38.3
70,000	32.5	30.5	48.5	43.3
90,000	36.0	34.0	48.5	49.3
125,000	40.6	40.9	53.5	65.5
200,000	48.1	51.4	63.5	70.8
Couple with two children				
50,000	13.4	4.5	30.7	32.3
70,000	20.2	12.9	43.5	38.3
90,000	25.4	18.7	43.5	43.3
125,000	31.5	27.5	48.5	60.5
200,000	40.8	42.1	63.5	70.8

Source: OECD Economic Surveys, Norway, 1983, p. 38.

[a]Includes employees' social security contributions.

When income taxes were cut, there was usually a compensatory increase in other taxes to protect government revenue. Transportation, telephone, postal services, liquor, and cigarettes were favorite targets. Despite the acknowledged need for tax reform, Labor remained wedded to high marginal tax rates for above-average income groups. In 1980, the minister of finance emphasized that "a progressive income tax remained an essential element of equalization policy. . . . A clear majority of the members of the Storting recognized the need for a high level of taxation to secure a basis for expansion in areas of common interest . . . income equalization was, and would remain, a key policy target."[4]

The labor market effects of the income tax policy were predictable. Flexibility was reduced by the unwillingness of wage earners to work overtime or on night shifts. In 1982, for example, a skilled metal worker -- and these were the ones in shortest supply -- would have earned about Kr. 110,000 working regular hours. Already faced with a marginal income tax rate of 50 percent if single and 43 percent if married with two children, he might have moved up to 60 percent if he were induced to work and earn more. Inordinate increases in nominal wages were required merely to keep real disposable income unchanged as a result of inflation and bracket creep.

Prior to 1970, the central government normally ran a budget surplus. This continued until 1975; after that year, the budget would have been in deficit every year had it not been for oil revenues (Table VI-3). Central government consumption and fixed capital formation changed little over the decade, so the increasing deficits were due to rising transfer payments.

Table VI-3. Central Government Budget Balances, Norway, 1970-1984[a]

(percent of GDP)

Year	Including Oil Sector	Excluding Oil Sector
1970	0.7	0.7
1971	3.2	3.5
1972	3.6	3.8
1973	5.4	5.9
1974	3.9	4.2
1975	3.4	2.7
1976	1.8	-0.1
1977	-0.4	-2.3
1978	-0.6	-3.9
1979	0.7	-4.1
1980	5.6	-3.3
1981	4.4	-4.7
1982	4.1	-5.0
1983[b]	3.9	-6.1
1984[b]	4.8	-6.0

Source: National Budget, various years; Norges Bank, Economic Bulletin, 1985/2, p. 119.

[a]Excluding loan transactions.
[b]Preliminary

The government fiscal system was increasingly strained due to the growing transfer payments and larger subsidies to industry and agriculture. The budgetary balances after 1976 were misleading when judging their impact on the level of demand. The budget deficits excluding oil revenues were expansionary; the oil revenues simply permitted a further increase in living standards, since the taxes from which the oil revenues were derived had little effect upon demand. The oil revenues were a free good as far as the tax-expenditure balance was concerned; they augmented the inflationary gap.

Faced with rising inflation, the Labor government found itself on the horns of a dilemma. It was being attacked for high income taxes that were only partly offset by periodic adjustments. High marginal tax rates reduced productive efficiency. Raising taxes on alcohol, tobacco, gasoline, and the like added fuel to the inflation. Bringing about a better balance between supply and demand by reducing government expenditures, on the other hand, ran counter to Labor's social goals.

Labor's first budget declaration in 1971 contained the statement that "in the years to come it would be necessary to increase public expenditures at a more rapid pace than the gross domestic product in order to solve the many tasks in the public sector."[5] A decade later, it was acknowledged in the National Budget that a tighter fiscal policy was desirable, but "on the other hand, there is a limit as to how much central government expenditure can be reduced without making it more difficult to maintain full employment in the short run. Importance is also attached to the desire to continue further high priority public sector tasks in a number of fields."[6]

When the conservatives came to power in 1981, they announced their intention to reduce both government spending and taxes. They planned to make a full adjustment of tax brackets to allow for inflation, to raise the tax allowance for children, and to reinstate the 80 percent maximum limit that Labor had removed. The tax burden on the self-employed and corporations was to be reduced. Charges for state-owned transportation, the postal service, and fuel were to be raised to reduce operating deficits.[7]

It is easier to create a welfare state than to reduce its scope. The first Conservative budget, that for 1983, "represented a compromise between the need for sustaining employment in the period ahead and the necessity of safeguarding employment in the longer term." Government expenditures were not to be cut, but merely to grow more slowly. The budget was to have a "clear social profile," concentrating assistance on those most needing it. While there was to be a gradual reduction of industrial subsidies, "in the present situation it was necessary to proceed more slowly in this respect."[8]

Foreign Borrowing

Before the oil revenues began to come in, the Labor government resorted to foreign borrowing to finance its budget deficits. It had no difficulty in finding willing foreign lenders. The domestic oil companies also borrowed heavily to provide capital for the investment that was needed.

It would have been difficult to finance oil investment out of domestic savings. Government borrowing to finance increased consumption in anticipation of oil revenues was a much more debatable policy. Norway's net external debt rose from 23.7 percent of GDP in 1975 to 43.8 percent in 1979. In 1975, government borrowing was about 25 percent of private borrowing; in 1979, it had risen to 70 percent. In the latter year the foreign debt/GDP ratio was one of the highest among OECD countries.

The borrowing spree came to an end in 1979 when the government began to pay off the debt. The net foreign debt fell from a peak of Kr. 103.6 billion at the end of 1979 to 75 billion at end 1984. However, by the latter date, most of the debt was attributable to the oil and shipping sectors, while central government debt had been largely paid off.[9]

Poor countries that discover oil are tempted to borrow from abroad in order to raise consumption levels, even before the oil flows. International agencies usually caution against this, urging that the proceeds of oil sales be invested in growth-producing assets. Given Norway's affluence in the pre-oil years, it is difficult to understand why the government borrowed from abroad to finance consumption when it was not faced with the tremendous pressures confronting developing countries, many of whose people live at the very margin of existence. There were certainly rising expectations in Norway, but they were hardly revolutionary.

Credit Policy

The Labor government's policies with regard to the allocation of credit were due in part to its ideological commitment to low interest rates. There was also a deep distrust of the private banking system despite the fact that it was subject to a degree of governmental regulation matched in few other democratic countries. The result was to deprive the government of a major weapon in the fight against inflation and to bring about a misallocation of investment resources.

Labor relied heavily on state-owned banks to distribute investment funds. Among them were specialized banks for housing, agriculture, industry, fishing, and education. It is important to note that all of these banks lent at concessional interest rates. The Housing Bank dominated throughout, though its share declined from 57 percent in 1971 to 44 percent in 1984.

At the peak of their activity in 1979, the state banks provided over 40 percent of all domestic credit resources, with about 30 percent coming from commercial and savings banks and the rest from insurance companies, pension funds, and loan associations. By 1984, the state bank share had fallen to 11 percent, while that of the commercial and savings banks had risen to 50 percent, as a consequence of the greater emphasis placed on the private sector by the Conservative government.

The Housing Bank devoted most of its funds to new dwellings, financing about 80 percent of them at the height of its operation. The original purpose of the Bank was to help low income people build and improve their homes, but the net result was that "a vast number of people who have resources to finance their housing" were also being subsidized, and that Housing Bank loans "no longer had a sufficient social profile."[10] For the state banks as a whole, the subsidy element in 1979 amounted to as much as 40 percent of the total loan amounts; the interest rates were lower than the government had to pay to raise the credit resources.[11]

The government also regulated the private credit market by strict control of interest rates and other conditions for many types of commercial bank loans and corporate bond issues. The OECD called the Norwegian credit market "one of the most thoroughly regulated among OECD countries."[12] It was the Ministry of Finance that formulated credit supply targets and decided on appropriate policy instruments; the Bank of Norway had little independent authority. Politics were in command.

The low interest rate policy severely hampered the monetary authorities in their quest for price stability. Government bonds were issued at interest rates well below real market levels. Private buyers were uninterested in them, and they were marketed by assigning compulsory quotas to private banking institutions. The type of open market operation that the Federal Reserve Bank of the United States employs in controlling the money supply was unavailable to the Bank of Norway.

A committee of experts appointed by the government in 1979 to review its credit policies advised some major changes. Petter Jakob Bjerve, the chairman of the committee, summarized its conclusions:

1. Interest rates on government bonds should be determined by market supply and demand, which would make open market operations possible. Market rates should also be charged for government loans to the state banks. Even if the latter were to lend at concessional rates, the government subsidy element would be a direct charge on the state budget, instead of being hidden.

2. Private credit institutions should also be permitted to charge higher interest rates. The committee did not go so far as to recommend that they should lend at market rates, but thought that the regulations should be more flexible.

3. The committee found that net debt increased with income, so that high income groups would lose and lower income groups gain (relatively) from increased interest rates.

The committee concluded that interest rates were not a suitable instrument for redistributing income.[13]

Leaders of the various political parties were asked their opinions of the committee report. Trygve Bratteli, who had been Labor prime minister until 1972, agreed with the recommendations, as did Kare Willoch of the Conservative Party and Lars Korvald of the Christian Poeple's Party. Johan Jakobsen of the Center Party expressed some reservations:

> An interest rate policy based on the proposition that the market -- supply

> and demand -- should in the main determine interest rates entails, in my opinion, the danger that important social tasks will not receive the necessary supply of capital. Who will find it difficult to compete in a market-determined capital market? There is ground to fear that small enterprises, municipalities that are financially weak, firms with low earnings, highly competitive industries, and young families which are just setting up their homes will be the most affected. This is why the Center Party is sceptical about eliminating interest rate controls and ideas about freeing interest rates.[14]

Hans H. Rossbach of the Liberal Party shared this view. The strongest opposition to the recommendations came from Hanna Kvanmo of the Left Socialist Party, upon which the Labor government was dependent for its majority. She urged that the entire banking system be nationalized and that all loans be made by the state banks. "Direct regulation of investment through credit policy is necessary in a mixed economy like that of Norway."

In response to the report as well as to the pressure of rising prices, interest rates were to be permitted to rise by one percent as of January 1, 1981. Housing subsidies were to be strictly limited to those without adequate financial means. While flexible interest rate regulation was approved in principle, "it was stated that the authorities must retain sufficient control of interest rates <u>to avoid a shift of power</u> from the public authorities to the credit institutions on matters of economic policy." [Emphasis mine] The state banks would be required to pay the same rates for the capital they acquired from the government as the government had to pay for the bonds it issued to eliminate hidden subsidies. However, interest rates that private banks could charge would continue to be determined by the Ministry of Finance.[15]

The conservative coalition that replaced Labor in 1981 took some steps in the direction of a market-determined credit system. The relative importance of the state banks was substantially reduced, as already noted. A market for short-term government securities was established to enable the Bank of Norway to engage in open-market operations as a means of controlling the money supply. The private banks were largely relieved of obligatory purchases of government bonds. The first new bank in 35 years opened its door in 1984, and foreign banks were to be given permission to enter Norway.

Interest rates have been permitted to rise to higher levels, although the Ministry of Finance still retains some degree of control. Since 1984, the money market rate of interest has been fixed at between 13 and 13½ percent, an increase of about 5 percent since 1979.

Government subsidies are difficult to terminate, even when there is a consensus that they distort the economy. The Labor Party is not alone in the view that low interest rates benefit low-income groups, despite considerable evidence to the contrary. During the 1970s, Labor had weakened the private banking system and pulled back only when it became clear that inflation and other unfavorable economic consequences might be the result. Suspicion and fear of private banks persist among socialists, even after they abandon the view that nationalization of industry is essential if the capitalist system is to be tamed.

Bank "Democratization"

Pursuant to its 1973 election program, the Labor government produced a white paper on "democratization" of the private commercial banks in order to provide greater community control. Several alternative proposals were advanced: permitting the bank to remain joint stock companies with the Storting to appoint a majority or a minority of their boards of directors; full nationalization of the banks; or transforming the banks into self-owned institutions with wide latitude in determining representation rights, essentially a half-way house between private and public status.[16]

A commission appointed by the government to examine these options rendered its report in 1976. It favored retention of the banks as independent institutions, but with public and employee representation on a governing body to be termed the "committee of representatives." It recommended that this committee be composed of between 18 and 45 members, one-ninth of whom would be elected by the employees. But agreement stopped there. The Labor minority favored selection of an additional five-ninths by the political authorities, while a minority advocated only three-ninths, leaving the shareholders with five-ninths control.

The board of directors would consist of between five and nine people, with the managing director a member _ex officio_. The remaining members would be chosen by the committee of representatives. It was assumed that the Storting and local authorities would select board members via their delegates on the committee of representatives. The latter would be the ultimate policy-making body of the bank, setting broad guidelines. The board of directors would retain authority over individual loan transactions. If the existing stockholders did not acquiesce in this arrangement, the government would buy their shares at market value.[17]

These recommendations produced a lively debate. The Norwegian Bankers' Association, pointing to the fact that the banks were already under stringent regulation, was strongly opposed to the idea that a committee on which stockholders were in a minority could decide, for example, that share capital should be increased or reduced, that the bank should be liquidated or merged, that dividends should be paid or not. It did not oppose the creation of committees of representatives, but insisted that stockholders should have a majority voice. Since some committee members would have little knowledge of financial matters, the functions of the committees would have to be carefully circumscribed.[18]

These and other objections were overridden by the Labor government, and on June 1, 1978, a modified version of the majority recommendations became law. Representative committees were to consist of 15, 30, or 45 members, of whom the Storting would elect eight-fifteenths on the basis of its proportional party strength. Stockholders would elect four-fifteenths, employees of the bank, three-fifteenths. Members of the board of directors would be chosen by the committee of representatives, except that one would be chosen by the bank employees. Within three years, stockholders could have their shares redeemed by the government at current market value.[19] It is interesting to note that during the two years this redemption option lasted, only a small fraction of the shares was tendered.

The Conservative Party promised that if it won the 1981 elections, it would restore the banks to stockholder control. This was done in 1982. However, the new governing structure was left intact except for the assignment of a majority vote to the stockholders. This was done in an attempt to avoid future controversy, but it did not succeed. The Labor Party, in its 1985 election program, committed itself to legislate equal representation for government, stockholders, and employees in order to ensure that social interests are better reflected in bank decisions. Now we shall have to wait until 1989 to see whether this promise is renewed.

It is difficult to find any practical reason that led the Labor Party to undertake this experiment in bank nationalization other than a desire to propitiate its left wing. The private banks were already under effective government control to an extent that made it difficult for them to carry out their normal functions. The state banks were allocating 40 percent of the credit resources available to the country, surely sufficient to allow the government to allocate investment along whatever lines it chose. There was no reason to believe that a banking system fully under government control would operate more efficiently than the semi-private system that prevailed. This episode must be looked upon as a triumph of ideology over common sense.

Price Policy

Deprived of the policy instruments that are normally available in a capitalist society to maintain economic stability -- reduction of government expenditures, control of the money supply, labor market discipline -- the Labor government resorted to direct price controls. Between 1972 and 1982, there were nine periods with some form of price control, sometimes as part of a broader incomes policy, other times in isolation.

Following the 1972 wage settlement and a liberal boost to farm incomes, prices began to rise, reaching a growth rate of 10 percent on an annual basis. Shortly before resigning due to its defeat on the Common Market issue, the Labor government imposed a price freeze until January 1, 1973. When this was lifted, prices resumed their upward course, and were moving up at a 9 percent annual rate in the summer of 1973. This time it was a non-Labor government that imposed selective controls, perhaps fearful of the impact of inflation on the forthcoming elections.

Despite the wage explosion of 1974, the new Labor government attempted to fight the consequent inflation by price subsidies, among other things. But this proved futile, and another price freeze came into effect during the final quarter of 1976.

Following a devaluation of the krone in February, 1978, prices were temporarily halted at their pre-devaluation level. This was followed in October by the most comprehensive control episode of the decade, during which all prices and incomes were frozen until the end of 1979. There were also controls on consumer credit: new car buyers had to pay 80 percent cash and the balance in six months, while for other consumer durables the minimum down payment was 50 percent, with the rest to paid within one year.

The immediate effect of the freeze was to halt the inflation. The price increase from 1978 to 1979 was the lowest since 1969. Only piece workers and those working under productivity bonus systems received higher earnings. The OECD reported "general agreement that the prices and incomes freeze has been instrumental in reducing price and wage increases."[20]

The lessons that should have been learned in the 1970s became perfectly obvious when controls were terminated at the end of 1979. Prices rose almost 11 percent in 1980. The government attempted to ride out the storm without new controls, but with prices rising at a 13.6 percent rate in 1981, the highest figure since the 1951 Korean crisis, another price freeze was instituted in August, to last until the end of the year. A national election campaign was in full swing, with inflation a major issue.

The Conservative government announced it would not resort to price controls. ". . . such regulations could prevent a necessary market adaptation and limit the growth of productivity, thereby increasing the price pressure in the longer run. Instead, the government will try to increase price competition and price and cost consciousness."[21] Although inflation remained at a high level in 1982, the Conservatives remained faithful to their commitment.

Apart from the inherent difficulty of freezing prices in a dynamic economy, the Labor government added to its difficulties by following a contradictory demand policy, as we shall see. There has been abundant experience from many countries that price freezes cannot in themselves solve the inflationary problem, but at best can provide an opportunity to adopt other remedial measures. The Labor Party failed to follow up with action that might have brought success. Even in the last national budget that it submitted to the Storting before it left office, the Labor government, while conceding that price regulation had negative effects on production and trade, warned that it might be necessary to introduce compulsory price reporting schemes in special areas if prices continued to rise.[22] There was still a lingering reluctance to let prices be determined by the forces of supply and demand.

Demand Policy

Labor's hallmark in the 1970s was adherence to an unrelenting Keynesian demand policy as long as there was any vestige of unemployment. The Keynesian economic model calls for budget surpluses and other instruments of demand restriction when the economy begins to overheat, but Labor appears not to have heeded this portion of it.

When the Conservatives held office briefly in 1972-1973, they announced that the growth of demand would be reduced in order to counteract rising prices.[23] Kleppe, when he became finance minister a year later, adopted a sharply different course. Fiscal policy was to be more expansionary in order to broaden the welfare state. He noted that other industrial nations were moving toward deflation, which could lead to unemployment and lower growth, "consequences which the [Labor] Government could not tolerate in the case of Norway."[24]

Exports contracted in 1975, but domestic demand rose by 6 percent, half contributed by oil activities. The following year, "faced with the problem of inflation and, by Norwegian standards, high unemployment (1.6 percent), the authorities opted in the autumn of 1975 for a strategy combining demand and employment-supporting measures with coordinated prices and incomes policies," leading to another 6 percent rise in consumption in 1976.[25]

The year 1977 was the fifth in a row in which the demand rise was considerably in excess of the output increase. It was not until 1978 that a change in demand policy produced a drop in domestic demand. This turnaround was motivated by the continuing inflation and a downward revision in oil revenue projections. Private consumption expenditures fell, and the rate of inflation moderated. The following year, 1979, was hardly a typical one in that a wage-price freeze was in effect. Nevertheless, fiscal policy tended to be stimulative, on balance.

Bjerve, in reviewing this record, concluded that while the goal of avoiding recession and unemployment had been achieved, there was a question whether some deviation from the demand policy followed might not have produced better results in the 1970s and provided a better basis for the economy in the 1980s. He concluded that "in 1975, 1977, and 1979, important policy actions were not taken prior to the elections in these years. Appropriate measures were adopted in 1976, which was not an election year, and in 1977 and 1979 after the elections. This raises the question of whether it is possible to conduct a suitable anti-cyclical policy if unpopular measures cannot be adopted during a large proportion of an election year."[26]

Labor was able to avoid economic recession by raising government expenditures, but this tended to create excess demand and inflation. The tight labor market engendered by the full employment policy added a wage-push component to other inflationary forces. The constraints that would normally have tempered this policy were weakened because of prospective and actual oil revenues. Insulated from the worst effects of the oil price increase, in part because of its heavy reliance on hydro power, Norway should have been able to do more than only marginally better in restraining consumer prices than its Scandinavian neighbors. From 1975 to 1981, Norway experienced a 9.6 percent annual rate of increase in consumer prices compared with 10.5 percent for Denmark and 10.6 percent for Sweden. What might be termed overfull employment was purchased at a high price.

CHAPTER VII

INDUSTRY AND FOREIGN TRADE

Norwegian manufacturing had a favorable record of growth after recovering from the ravages of World War II. Prior to 1974, the rate of growth of manufacturing output was approximately equal to the average OECD rate. During these years, Labor governments generally did little to interfere with the private manufacturing sector.

There were a few exceptions: a small fund was set up in 1962 to help firms adjust to changing market conditions. An export guarantee fund was created in 1960, and a fund to assist handicrafts and small firms in 1967. Investment resources were available through the banking system; people worked hard; industry prospered.

A good deal of this industry was based upon the cheap electric power afforded by harnessing the country's many waterfalls. Among the important energy-intensive industries were the refining of aluminum from imported bauxite and the production of chemical fertilizers. Aluminum output increased from 47,000 tons in 1950 to 522,000 in 1970, fertilizer output from 42 million kilograms to 839 million for the same two years. Industries based on Norway's forests -- timber, pulp, paper -- also expanded rapidly.

There has been some recent criticism of Labor's historical policy of encouraging industry based on natural rather than human resources. The argument advanced is that the export of large quantities of semi-fabricated commodities rendered the economy too vulnerable to swings in the international business cycle and resulted in a low value-added component. "Easy access to cheap energy can in many ways be seen as a curse for Norwegian industry. It provided an alternative to the development of knowledge based industry." Sweden is presented as a better development model.[1]

Such criticism flies in the face of the economic rule for the allocation of productive resources: invest in industries in which the cheaper factors of production dominate. With its small internal market and scarce manpower resources, a more diversified and labor-intensive industry would have led to a slower rate of growth. It is not quite clear what knowledge-based industries could have been created early, or indeed, precisely what the concept means. If it involves research and development, that is an expensive activity in which Norway had no particular advantage. In fact, market considerations were the primary determinant of industrial development, and attempts to steer a different course could have been counter-productive.

The large processing industries were by no means the only ones in the country. There were, and still are, a considerable number of enterprises producing textiles, furniture and other wood products, and metal products, that had the advantage of proximity to the domestic market. But they were mainly small; the average employment per plant in 1970 was 37 for textiles, 13 for furniture, 10 for other wood products, and 22 for metal products. By comparison, the capital-intensive chemical plants employed an average of 153 persons.[2]

These smaller enterprises developed with little government support. Some failed, others arose to replace them. They were important in providing both employment and output. In 1981, firms with less than 50 employees constituted 90 percent of all manufacturing enterprises in the country, produced 27 percent of total value of output, and provided 31 percent of manufacturing employment.[3] A government commission appointed in 1979 and chaired by Finn Lied made the following observation:

> It is strongly asserted by the management of small firms and by their organizations that fewer and fewer individuals are willing to take on the burden of creating new enterprises. This is due to the fact that the incentives emanating from social recognition and economic profitability are relatively weak in today's society.[4]

The Performance of Manufacturing

Beginning in 1974, Norwegian manufacturing began to stagnate. This emerges clearly from the data in Table VII-1. In 1984, manufacturing output was still below the 1974 peak. In this respect, Norway appears to have done worse than most OECD countries. From 1972 to 1982, the average annual increase in real value added by manufacturing was 0.3 percent for Norway compared with 0.6 percent for Sweden, 2.2 percent for Denmark, and 2.2 percent for the smaller countries taken as a group.[5]

Table VII-1. Indices of Manufacturing Performance in Norway, 1970-1984

(1970 = 100)

Year	(1) Manu-facturing Output	(2) Hourly Earnings of Men	(3) Employment	(4) Annual Output per Employee	(5) Unit Labor Costs
1970	100.0	100.0	100.0	100.0	100.0
1971	103.5	112.4	n.a.	n.a.	108.0
1972	107.3	122.3	102.3	104.9	114.2
1973	112.7	135.3	103.2	109.2	120.8
1974	116.9	158.8	103.7	112.7	137.1
1975	114.2	190.2	108.7	105.1	166.9
1976	114.2	221.4	110.1	103.7	187.8
1977	113.1	245.6	108.2	104.5	205.1
1978	110.8	265.0	104.8	105.7	221.5
1979	113.1	272.5	101.6	111.3	221.8
1980	114.2	298.0	102.6	111.3	239.5
1981	113.1	328.3	103.2	109.6	261.1
1982	110.8	361.9	101.6	109.1	268.9
1983	109.6	394.8	94.2	116.3	282.6
1984	113.1	424.4	96.1	117.7	296.7

Sources: Col. (1) Central Bureau of Statistics, _Historical Statistics_, 1978; National Budget, 1985.
Col. (2) Central Bureau of Statistics, _Statistical Yearbook_, various issues.
Col. (3) OECD, _Main Economic Indicators_, various issues.
Col. (4) Col (1) ÷ Col. (3)
Col. (5) OECD, _Historical Statistics_, 1960-1982; _National Budget_, 1984-85, p. 14.

Other countries were able to achieve substantial increases in output despite declining employment by raising productivity, but the Norwegian productivity record was not favorable. The OECD summarized Norway's productivity in the following terms:

> While the slowdown in measured labor productivity growth has been a pervasive feature of OECD economies since the early 1970s, Norway has been among the most severely

> affected. Norwegian industry's relative productivity performance deteriorated by 15 percent over the past decade. The relative decline was particularly pronounced in the four years to 1977, followed by virtual stability thereafter. Productivity growth (measured as value-added per man-year) decelerated in all industrial branches with the sharpest slowdown recorded in the export competing industries.[6]

Various explanations of this phenomenon have been advanced, among them the following:

1. Labor costs. The wages demanded by trade unions, and agreed to by employers, have resulted in a large increase in unit labor costs. The rate of increase was particularly great from 1974 to 1976 as a result of the 1974 wage explosion. Unit labor costs rose by almost 40 percent between these two years.

If nominal wages do not increase more rapidly than labor productivity, there is a good basis for the avoidance of inflation. How far from this standard Norwegian manufacturing deviated can be seen by comparing columns (2) and (4) in Table VII-1. Nominal wages more than quadrupled from 1970 to 1984 whereas productivity was up only 18 percent.

Nominal wages did not increase more rapidly in Norway than in its major trading partners, as Table VII-2 indicates. However, nonwage payroll costs rose in Norway from 30 percent of wages in 1970 to 40 percent in 1979, which must have accounted for some of the discrepancy between the unit labor cost and the earnings/productivity data. In any event, by 1983, unit labor costs in Norway were 28 percent higher relative to those in its major trading parners compared with 1970. There can be no doubt that this was a major factor behind the difficulties faced by Norwegian producers.

2. Output and investment. According to the so-called Verdoorn effect, there is a positive correlation between manufacturing output and productivity. New capital investment, by providing modern equipment, contributes to output via higher productivity as well as directly through a rise in the capital stock.

Table VII-2. Indices of the International Competitiveness of Norwegian Manufacturing, 1970-1982[a]

(1970 = 100)

Year	Relative Hourly Earnings	Relative Hourly Productivity	Relative Unit Labor Costs	Effective Krone Rate
1970	100.0	100.0	100.0	100.0
1971	101.7	99.7	101.6	99.3
1972	101.3	98.3	102.2	99.1
1973	100.0	96.1	107.1	103.8
1974	97.2	97.3	111.1	110.3
1975	98.5	94.4	120.1	113.9
1976	101.6	92.1	129.4	115.8
1977	103.5	89.9	135.6	116.2
1978	103.2	87.4	127.5	106.9
1979	97.3	86.4	119.7	106.9
1980	94.8	84.6	122.1	107.9
1981	98.7	85.1	125.0	106.9
1982	99.2	82.9	131.7	107.6
1983	102.0	82.3	127.8	102.4

Source: Landsorganisasjonen og Norsk Industriforbund, Norge som Industrinasjon, 1984, p. 48.

[a]Comparisons are with Norway's major trading partners, weighted by their share of Norwegian exports. The figures represent the Norwegian data relative to those of their trading partners.

Norwegian investment has remained high throughout the postwar years, including the 1970s. However, the proportion of total investment allocated to manufacturing fell from 15.2 percent in 1970 to 10.8 percent in 1982. Part of this decline was due to investment in oil, but even if the oil sector is excluded, the manufacturing share was down to 12.5 percent in 1982. Moreover, by the late 1970s, about 10 percent of industrial investment was allocated by law to labor and environmental protection, which does not contribute to productivity, at least in the short run.

Reducing the effectiveness of investment, and thus raising the capital/output ratio, may have been due to the fact that a substantial proportion of investment was allocated by the government on the basis of considerations other than profitability. This means that new capital was not being applied where it would maximize output, and that therefore more of it would be required to achieve a given level of output. Artificially low interest rates contributed to the scarcity of private risk capital and increased industry's reliance on government for investment funds.

3. Subsidies. Labor's policy of subsidizing entire branches of industry as well as individual firms contributed to the poor performance of manufacturing. This may seem paradoxical, but it is nonetheless true. The rationale for this policy was set forth by the Minister of Industry in a report to the Storting in 1980:

> During the sixties the principal purpose of industrial policy was to promote even growth. Since the middle of the seventies industrial policy has had a more defensive character with particular emphasis on maintaining the existing places of work and industrial plants. Industrial policy came to play an important role in maintaining full employment. Its main instruments were short run, countercyclical measures concentrated on enterprises that were encountering difficulties.[7]

Direct subsidies were in addition to such other means of support as tax relief, guarantees, protection against imports, and credit at concessional rates. The peak year was 1977, when subsidies amounted to 5.7 percent of value added in manufacturing. The purpose of the subsidies was to shield regions that were in economic difficulty and to protect jobs in individual plants and locations. Following are the major programs:

a. The Industrial Fund employed a wide spectrum of financial measures to assist manufacturing. These included aid to new enterprises, assistance in effectuating structural change, support to small enterprises to raise their capacities, purchase and introduction of new technology, and refinancing firms that found themselves in economic difficulty. Assistance took various forms: guarantees, underwriting stock issues, loans. While the fund was under the administrative control of the ministry of industry, it operated independently except when large amounts were involved.

The fund was conceived of as an "active institution to take the initiative in improving the long run position of Norwegian industry."

b. Liquidity loans were introduced in 1972 to provide short term financial relief for firms affected by cyclical downswings. Between 1972 and 1978, some 1,200 loans were made to 1,000 enterprises with about 100,000 employees, a quarter of manufacturing employment. Concessional interest loans to finance inventories were first used in 1975 in order to help even out production over the business cycle. From 1975 to mid-1977, some 79 firms with 47,000 employees were the recipients of such loans; this scheme was terminated as a separate program in 1978.[8]

c. The program to support specific branches of industry was designed primarily to assist labor-intensive industries, where failure would have had a strong impact on employment. Products affected were textiles, clothing, shoes, luggage, leather, furniture, doors, windows, glass and ceramics, and canning. Assistance was given in the form of grants to industry committees which used them for planning and allocations to individual firms. The goal was to stimulate cooperative endeavors to meet foreign competition.

A large proportion of these funds went to the shipbuilding industry, which had been hard hit by the worldwide decline in the demand for ships. Subsidies took the form of low interest loans, reduced prices for work done for Norwegian shippers, and employment support grants. The government also guaranteed long term loans. Over the two years 1977 and 1978, shipbuilding contracts secured by Norwegian yards as a result of the subsidy program amounted to 145 percent of the value of shipbuilding output in 1976. About 20,000 people were employed in the new construction part of the industry, so that the subsidies guaranteed about 30,000 jobs over the two years. The ministry of commerce estimated that the annual cost per job was over kr. 100,000, at a time when average annual earnings in manufacturing were on the order of kr. 70,000.[9]

There had been a report by a government commission appointed in 1976 to survey this ailing industry and had come up with recommendations for its cure. The committee majority felt that it was necessary to keep the yards in operation until European orders picked up, and recommended that capacity be maintained at about 14,000 employees, a 40 percent reduction from 1975. A minority of the committee favored a reduction to 8,000 jobs.[10] As things turned out, the subsidies exceeded even the majority recommendation; more than 50 shipyards were kept going in whole or in part by government largesse.

The 1980 National Budget promised to terminate shipbuilding subsidies in 1982, on the theory that competing countries would also eliminate their support schemes. However, the Conservative government continued the practice, and in 1984, new ship contracts received subsidies amounting to 27½ percent of the contract price in order to encourage Norwegian shipowners to place their orders in Norway.

There was a change in the character of shipbuilding subsidies over the years. In 1977, orders placed in Norwegian shipyards were given a direct price subsidy of 20 percent, and $100 million in aid was provided to developing countries to enable them to buy Norwegian ships. The following year, the general price subsidies were reduced to 12 to 15 percent, while yards received subsidies of 10 percent for any ship bought by Norwegian owners. Beginning in 1981, the support largely took the form of low interest rates for long term financing.

Another major recipient of government support were firms that were taken over by the government as an alternative to bankruptcy. Among them were large enterprises in coal mining, pulp, electronics, and aluminum. They were often purchased at above-market prices, resulting in substantial losses to the government and diverting capital from more profitable areas.

d. Training subsidies were given to individual firms to prevent layoffs. Through 1979, 300 firms employing 30,000 had received such grants.

e. Government guarantees up to half of the cost were available to exporters bidding on contracts with less developed countries as well as those with state trading monopolies.

f. Regional subsidies were established to help maintain the existing pattern of population distribution. Agriculture and fishing were major beneficiaries, but industry also benefited from direct transfers, special tax abatements, government loans, and guarantees. Tor Halvorsen, the chairman of the trade unions, indicated his awareness of some of the pitfalls of this scheme:

> A strong emphasis on the goal of decentralized dwelling and more households in which both married partners work outside the home make for a strong demand to prevent geographical mobility. But if we are too strongly committed to the retention of the existing structure of plants and industries, the dwelling pattern may be more difficult to achieve. To put it another way, enterprise and industrial mobility may be a necessary condition for maintaining geographical stability.

He noted that in some recent closures of large factories, employees were able to secure new jobs within reasonable distance of their homes. While insisting that the LO opposed abandoning the possibility of assisting firms that were important to local labor markets, he did concede that "the goal of full employment does not mean that everyone shall be employed in the same job permanently," and that "if it costs considerably more to keep people in their present jobs than to let them take another, it must be expected that society will react."[11]

There was other ad hoc support as well. The payroll of the textile industry was subsidized by a sum equalling 10 percent of average hourly wages; some export industries were insured against loss from foreign exchange fluctuations; and the government acted as partial investor in some new enterprises, and took over some ailing firms, as already noted. Among them was the Tandberg electronics company, one of the most advanced technological firms in the country, which was taken over by the state in 1978, only to go into bankruptcy a year later.[12]

A survey conducted in 1982 indicated that about 16 percent of the manufacturing labor force was employed in industries in which the wage bill exceeded value added. Over 50,000 jobs were in this category. The question is often asked -- and not only in Norway -- "what is wrong with keeping people working with government support as an alternative to the payment of unemployment benefits? At least there is some product, and less psychological harm is suffered by the individual who might have been laid off."

Per Kleppe, one of the chief architects of the subsidy

program, conceded that "while the main objective of subsidies was to sustain employment in the short run . . . this could possibly have resulted in a postponement of essential rationalization and restructuring."[13] This was an understatement. As the OECD put it, "with a few exceptions, the subsidies and loans granted were not conditional on steps being taken by the enterprise concerned towards rationalization, product development, or structural change."[14] Grants and loans tied manpower and other productive resources to firms from which they should have been released to more profitable enterprises. Labor hoarding became widespread, putting pressure on wages and prices. The director of the Bank of Norway stated: "The explanation for the weak trend in productivity is in large measure the fact that Norwegian companies retain workers who are not fully employed. Public support measures have often made this possible."[15]

Labor's commitment to the maintenance of full employment at any cost left Norwegian manufacturing in a weaker position than it would have been had the government refrained from intervention on so large a scale. In the long run, this critical sector of the economy would have been able to supply more rather than fewer jobs.

The present government of Norway has recognized the necessity of altering subsidy policy so that restructuring is not hindered. But it is difficult for any Norwegian government to countenance a rash of plant closures, particularly when significant local interests are involved. Industrial subsidies in 1984 were actually above the 1981 level. The lesson that may be learned from the Norwegian experience is that once in place, government support to industry is very difficult to terminate because of political pressures to continue it.

4. Entrepreneurial failure. A recent book by Øystein Noreng to which reference was made above places the major blame for the weakness of Norwegian industry on the inability of entrepreneurs to exercise sufficient influence on the direction of industrial growth. "The Norwegian capitalist class has been weakened because of its dependence on foreign capital and because of regional conflicts of interest. Here there is a clear difference from other Nordic countries. The contrast is particularly sharp between weak and fragmented Norwegian capitalism and concentrated Swedish capitalism, with long experience in international industrial activity and a leading position in important areas."[16]

The argument continues: "Norwegian capitalists are less able to confront their employees than their counterparts in any European country. They retain a smaller share of profits. A great many plants are small, often family-owned, and capital markets have less importance than in other developed countries." He feels that the problem lies in the structure of Norwegian industry, and more in unsatisfactory economic and administrative management than in a low level of technical knowledge. Moreover, corporate boards of directors normally play a passive role. They are heavily weighted with lawyers, who merely ratify management decisions.

Equality and stability are more highly valued in Norway than differences and change. As for the influence of the left:

> In contemporary Europe Norway is outstanding in that left socialist forces and politically radical intellectuals are more eager to protect a traditional social model than to further change and economic development. An intense desire to preserve, with a strong mixture of nationalistic romanticism without practical sense, imparts to portions of Norwegian left socialism a comically reactionary stamp. The industrial program of the Socialist Left party, with its predilection for increased agricultural employment and opposition to robots, is a good example.[17]

The author calls Norway the world's most organized democratic society. Policy is determined by special interest groups; the market mechanism has been put out of play. Collective decisions dominate. Economic adjustment is often more a matter of political influence than of costs and earnings. The solution, he feels, is a combination of government planning and market competition on the "Japanese model." The government ought to stimulate competition in new industries, and should be heavily involved in research and development. A pure market solution would run counter to the traditions, values, and power structure of Norwegian society. It should be the function of government to allocate investment resources to priority industries, particularly those that generate a high value added.

These observations are very interesting, but counter-arguments can be made. Until the 1970s, Norwegian manufacturing was doing quite well. From 1960 to 1973, real value added in this sector rose at an annual rate of 4.6 percent compared with 4.9 percent for all OECD countries. Productivity was at the low end of the OECD scale, but profitability was not inordinately low. From 1968 to 1973, for example, operating surplus as a percentage of value added in manufacturing was higher in Norway than in the United States.

Norwegian entrepreneurs in shipping were among the best in the world. Norway's major export industries appear to have been well run, on the whole. Nor is it necessarily true that small firms are economically handicapped. In 1981, there were 873,000 manufacturing enterprises in Japan employing 12.9 million workers, or 14.8 per establishment. Recent American experience suggests that the bureaucratic managerial structure of large firms is less innovative than the smaller firms that are pioneering new technologies. Some of the largest firms in the United States -- U.S. Steel and International Harvester, to name two -- could match any Norwegian firm in managerial incompetence.

Moreover, in a small economy with a widely scattered population it was inevitable that many small firms catering to local markets should appear. The historical structure of Norwegian manufacturing does not seem to have been the basic problem; it was, rather, its failure to respond to market signals because the government relieved it of the necessity of doing so.

The question may be asked whether vigorous and effective entrepreneurial activity is compatible with a socialist-run government, even of a mild social democratic variety. Faced with high tax rates, profit limitations, and a general hostility to the private sector, there is a reduced incentive for managers to innovate. This, rather than enterprise size or value-added ratios, is more likely to be the industrial problem of the future in Norway, and in many countries of Western Europe as well.

Foreign Trade

Between 1970 and 1982 there was a steady decline in Norway's share of the foreign markets for its traditional goods. Moreover, portions of the domestic market were lost to imports from abroad. The extent to which this occurred is estimated in Table VII-3. Over this period of twelve years, Norway lost more than 30 percent of its foreign markets and 15 percent of its domestic market to foreign competitors.

Table VII-3. Indices of Export Market Shares and Import Penetration for Norway, 1970-1983[a]

(1970 = 100)

Year	Export Market Share	Import Penetration
1970	100.0	100.0
1973	97.0	106.0
1976	84.4	104.8
1979	75.8	110.0
1980	73.8	112.7
1981	71.9	115.3
1982	69.3	115.3
1983	73.0	122.3

Source: OECD, Economic Surveys, Norway, 1985, p. 10.

[a]Calculated in terms of volume. Import penetration is the ratio of value of imports of traditional goods to value added in manufacturing plus imports of traditional goods less exports of traditional goods. Oil is excluded.

The OECD estimated that about one-third of the foreign market loss was due to two factors: the failure to adjust the commodity composition of exports in line with foreign demand and the relatively slow economic growth of the principal foreign markets (the United Kingdom and Scandinavia customarily absorbed half of Norwegian exports). The remaining two-thirds is attributed to reduced price competitiveness and to such factors as product quality and inability to meet delivery dates. In addition, the reduction of profit margins after 1975 may have induced many producers to abandon the export market.[18]

If there was one factor behind the loss in competitiveness of Norwegian industry, it was lagging productivity. The data in Table VII-2 put its relative loss at 17 percent for the period 1970-1982.

The decision by the Norwegian government to maintain its currency at a high international value also proved costly for exports. The krone was revalued upward in

1973 contrary to the advice of a majority of the board of directors of the Bank of Norway. The purpose, in addition to reducing the pressure on the domestic price level, was to curtail export sector earnings in order to modify forthcoming wage demands. The decision to join the European monetary cooperation system (the "snake") led to continued appreciation of the krone, which by 1977 was 16 percent above its 1970 value. The message that this was hampering exports finally got through, and in 1978 there was a devaluation of 8 percent. A subsequent tendency toward appreciation was countered by further devaluations in 1982 and 1984.[19] Fiscal and monetary restraint would have been a more appropriate instrument than a high krone value in the fight against inflation, but this would have come into conflict with Labor's policy goals of full employment and expansion of the welfare state.

There were differences in the extent to which the general decline in foreign trade affected various industries, but only a few of them were able to recover the market shares they lost. Clothing, shoes, fur and leather, and furniture were particularly hard hit; they were labor intensive and have suffered in other industrialized countries as well from the cheaper products of the less developed countries. The mistake in Norway, as elsewhere, was in trying to save them by subsidies and protective measures. The industries that did recover were those with high energy and low labor inputs.

The Oil Factor

In one sense, the decline of Norwegian manufacturing can be attributed to the discovery of oil. Norway traditionally ran a deficit in its balance of trade, which was offset by earnings from shipping. In 1970, for example, the trade balance deficit was kr. 8,137 million, while freight earnings were kr. 7,790 million. The trade deficit reached kr. 23,255 million in 1977, and since shipping earnings had declined and the oil had just begun to flow, Norway had a huge current account deficit. By 1982, although the commodity trade deficit was still large, petroleum exports more than covered the deficit.

Had it not been for oil, the policies that weakened manufacturing could not have been effectuated. With shipping in eclipse, the foreign balance constraint would have necessitated an increase in exports or a decrease in imports, or both. Wage restraint and measures to raise productivity would have been essential. Consumption would have had to be reduced.

Too many duties were assigned to manufacturing. Not only was it to produce goods, but it was also to provide "meaningful work" for those it employed, and to be administered "in democratic ways." It was to preserve the prevailing population distribution model. The environment was not to be disturbed. Manufacturing was to "shape greater equality in the working life choices for men and women."[20] But these tasks were contradictory in many ways, and the ones that were "social" took priority over the "productive" ones.

Oil money was used to preserve the existing pattern of industry. The restructuring necessary to meet changing market demands was slowed, if not stopped. New initiatives were not encouraged. As Noreng put it, "the market became less important in the Norwegian oil economy, but the role of the bureaucrats became more important."[21]

Would the Norwegian people have been better off had there been no oil? Living standards would certainly not have risen as much as they did during the 1970s. Unemployment would have been greater, there would have been more disruption of residence patterns. On the other hand, a more productive and efficient manufacturing sector might have emerged, with the capacity to provide more remunerative and "meaningful" jobs. Not even unemployment is as demoralizing as slack work which is perceived to be unproductive; the sociologists have established that. In the longer run, manufacturing might have been better equipped to provide the engine for economic growth, as it had done in the past.

Had different policies been followed, the oil revenues could have enhanced both the productive base of the economy and the immediate quality of life. That would have required either a wiser governing elite or the willingness to tolerate greater market freedom. What was called for was more humility in deciding whether the state's decisions were apt to be superior to those of individuals.

To quote Noreng once more: "After the Storting election in the fall of 1973, a diminished Labor Party was able to form a government, but it faced a strong left socialist challenge. In those years, radical ideas dominated Norwegian political debate much more strongly than at any time after 1945."[22] Whatever the political reasons for the industrial policy of the Norwegian Labor

Party during the 1970s, the economic results are clear: Norway emerged from a decade of new wealth with a weaker industrial base than it had enjoyed when it had been obliged to rely on the skill and hard work of its people, operating in a regime of limited natural resources.

CHAPTER VIII

THE WELFARE STATE

Every modern democratic, industrialized nation can be characterized as one in which the welfare of the people in such matters as social security, health and education, housing, and working conditions is the responsibility of the government. But there are great differences among nations in the degree to which this responsibility is acknowledged and exercised. In some countries the government may be the major and even the sole provider of the goods and services that enter into individual welfare; in others, it may play a role secondary to that of the private sector and the individual himself.

An index commonly used to determine where in the welfare spectrum a particular country belongs is the proportion of the national product that the government takes in through various forms of taxation. Not all of this is necessarily redistributed to citizens for welfare purposes. A portion will be allocated to defense and to other governmental functions outside the "social" rubric. However, the more the government preempts, the greater is the potential for welfare expenditures.

By this measure, Norway belongs to a small group of welfare super-states, a position it already enjoyed in 1970. In recent years, in only four OECD countries, the three Scandinavian nations and the Netherlands, did government levies account for more than half of the GDP.

It is more difficult to categorize countries by looking at government expenditure data. Norway, for example, is only slightly above the OECD average in the ratio of social security expenditures to GDP, well below France, the Netherlands, and Sweden. But gross expenditures for old age pensions, which are normally the largest category of social security expenditures, do not necessarily indicate how well pensioners are treated. International variations in the age of retirement, the age structure of the population, the coexistence of private pension schemes, and services provided directly to the aged render it difficult to base comparisons on expenditure data alone.

There are also differences among countries in the emphasis which the various welfare programs receive. This is illustrated by the data in Table VIII-1. Norway's

Table VIII-1. Allocation of Benefits by Social Security Program, 1980. Selected Countries

(percent of total benefit expenditures)

Country	Medical Care	Employment Injuries	Pensions	Unemployment	Family Allowances
Norway	47.9	0.2	43.9	2.2	5.8
Denmark	36.0	0.8	42.0	16.9	4.3
Sweden	48.9	1.0	41.4	1.7	7.0
Netherlands	34.3	---	51.2	5.0	9.5
Germany	31.6	3.3	51.5	7.4	6.2
United Kingdom	38.9	1.4	44.4	3.3	12.0
United States	29.8	7.0	55.1	8.1	----
Japan	53.7	3.6	35.4	6.0	1.3

Source: International Labor Office, The Cost of Social Security, 1978-1980, Table 8.

expenditures for unemployment benefits were low because of the low level of unemployment that prevailed. A greater proportion of expenditures went into medical care than for any country except Sweden and Japan; the latter provided little in the way of other benefit programs. The low relative share Norway allocates to pensioners does not mean that its pensioners have fared badly. On the contrary, they do better in absolute terms than the aged in almost any other country.

It is safe to say that Norway has one of the most generous social security systems in the world. There is scarcely a human contingency not covered by a government support program. No Norwegian is living in poverty except by his own choice. This is a remarkable achievement, and the Labor Party, which was the principal architect of the system, deserves full credit for creating it. No political party advocates its dismantlement. But as with most things, there are costs involved that are not always appreciated, and it is fair to ask whether the system has gone too far.

By 1970, all the major social security programs had already been established in Norway. During the 1970s they were liberalized. Social expenditures rose as a proportion of GDP from 14.7 percent in 1970 to 21.7 percent in 1981.

To pay for expansion of the programs, taxes had to be raised. Table VIII-2 shows the proportion of tax revenues that were necessary to finance the social security system. It illustrates the importance of oil revenues in permitting welfare payments to rise. Had it not been for the oil, financing the national insurance system would have required one-third more of total tax revenues in 1982 than in 1970, that is, oil taxes provided one-third of the increase in social security costs.

The growth of industrial fringe benefits, which were largely of a "social" character, is shown in Table VIII-3. The 35 percent increase of these benefits over the decade contributed to Norway's rising unit labor costs.

Social Programs

A brief description of the major social programs and how they expanded during the 1970s will serve to clarify the meaning of the welfare state as it has developed in Norway.

Table VIII-2. National Insurance Income as a Proportion of Total Tax Revenue, Norway, 1975-1982[a]

(percent)

Year	Total Tax Revenue	Tax Revenue Excluding Oil Taxes
1970	26.5	26.5
1972	19.0	29.0
1974	31.2	31.3
1976	29.3	30.5
1978	28.8	30.6
1980	26.8	31.9
1982	29.5	35.2

Source: NOU 1984:10, Trygdefinansiering, p. 52

[a]The National Insurance system does not administer all social programs. In 1980, for example, National Insurance expenditures constituted only 72 percent of all government expenditures characterized as "social." Thus for 1980, National Insurance income and other social expenditures would have taken 37 percent of total tax revenue.

Table VIII-3. Non-Wage Labor Costs in Industry, Norway, 1970 and 1981

(percent of hourly wages)

	1970	1981
Payment for days not worked	14.2	20.2
of which		
Paid legal holidays	3.5	3.5
Vacation payments	9.8	10.1
Payments due to illness	0.9	6.6
Social security contributions	15.6	21.2
Other expenditures	5.7	6.7
Total	35.5	48.1

Source: OECD, Economic Surveys, Norway, 1982, p. 39.

A national pension scheme was established in 1936, and by 1970, coverage was universal. It includes disability as well as old age pensions. There is a basic pension amount which increases with years of coverage. In case of married couples, if both are receiving pensions, each is entitled to 75 percent of the base, which is adjusted annually to compensate for increases in prices and incomes. There is a special supplement for pensioners with dependent spouses who do not receive pensions.

An income related supplement to the base amount was introduced in 1967. Pension point credits are awarded to a wage earner for each year of coverage in which earnings exceed the current base amount. A special supplement was added in 1970 for those who did not qualify for the regular supplement, i.e., low wage earners. In 1970, the average supplement was 7.5 percent of the base for a single person; by 1982, it had risen to 49 percent.

This and other factors have led to a sharp increase in pension costs. The pensionable age was reduced from 70 to 65 years in 1973. A special supplement was added in 1970 to compensate for the levying of a value added tax. The population has been growing older, so that the number of old age pensioners rose from 441 thousand in 1973 to 552 thousand in 1983. Pension point credits automatically led to higher pensions as real earnings rose.

There are disability pensions and a survivor benefit program for children under the age of 18 both of whose parents are dead; the age of support rises to 20 if the individual is attending school. Surviving spouses are also entitled to benefits for child support and education.

The cost of the pension system is shown in Table VIII-4. The large jump from 1972 to 1974 was due in part to the lowering of the pensionable age. The relative decline after 1978 was a consequence of the wage-price freeze promulgated in that year. Minimum pensions continued to rise, but higher pensions were held back.

The goal of the pension system was not to provide retirement income proportional to prior earnings, but rather to give lower income groups a higher proportion of pre-retirement income. As things worked out, an individual retiring in 1983 with an income of kr. 80,000 in the last year of his working life (roughly that of an average blue collar worker in manufacturing) would receive a pension equal to 72 percent of disposable pre-retirement income. The comparable figure for a person earning kr. 125,000 (for example, a supervisor in industry) was 63 percent, and for a chief engineer earning kr. 250,000, 50 percent.[1] It must be remembered that private pension schemes are not widespread in Norway, but at least for blue collar workers, the pension amount, adjusted annually for price changes, provides a high degree of financial security in old age.

Table VIII-4. Pension Costs and Benefits, Norway, 1970-1981

	Pension cost as a percentage of GDP	Index of ratio of pensions of a married couple to the average industrial wage for men (1970 = 100)
1970	5.1	100.0
1972	6.9	104.7
1974	8.9	99.0
1976	9.1	98.3
1978	9.6	115.4
1980	8.6	124.4
1981	8.5	129.6
1983	n.a.	134.6

Source: Stein Kuhnle, Velferdstaten, Oslo, 1983, p. 67; Rikstrygdeverket, Arsmelding, 1984-85, p. 154.

The tendency of pensions to outrun wages arose as a problem in the 1980s, as Table VIII-4 shows. The goal established in 1973 was to have the disposable income of pensioners and wage earners rise at the same rate. The basic pension was raised sharply in 1977 in order to compensate for the pension lag after 1972, but the government overshot the mark.

The health services provided by the state in Norway are virtually all-embracing. As in other countries, the costs have mounted rapidly, requiring a rising share of the national product. Basic medical coverage includes care by a physician, with the patient paying the difference between the fees charged and those fixed by the insurance system for the first two visits of the same illness, with no charge for further visits that may be necessary. Physical therapy, costly medicines, psychiatric treatment, and the provision of hearing apparatuses and orthopedic devices are also among the services covered.

Maternity costs are fully covered, and a working woman is entitled to 108 days off with pay after giving birth. If she returns in a shorter time, maternity pay can be transferred to the father if he undertakes care of the child. Non-working mothers receive lump-sum maternity benefits. In the event that a child under 10 years of age is ill, an employed parent has the right to sick pay for absence to care for the child up to 10 days a year in general, and in some cases up to 20 days. Travel expenses necessary for diagnostic procedures and treatment are covered, including overnight lodging if necessary.

Household assistance may be provided in case of long term illness unless a family member is available to help. People suffering from serious chronic diseases are entitled not only to free medical care, but also to payment for such extra expenses as transportation, special diets, telephone, and laundry. Some costs may be covered at the discretion of the Health Directorate, including treatment abroad for rheumatism and psoriasis, eyeglasses and contact lenses for children under 15 years of age, dialysis, exercise bicycles, respirators, and complete dental care.[2]

Employees have the right to sick pay beginning with the first day of illness, at full pay. The cost of the first two weeks of sick leave are borne by the employer, and thereafter by National Insurance. Self-employed individuals are entitled to daily sick pay equal to 65 percent of their previous income beginning with the 15th day of illness. Full coverage may be secured by paying extra premia to National Insurance. Sick pay is taxable and counts toward pension rights. It is payable for every day except Sunday up to a maximum of 312 days for the same illness.

Prior to July 1, 1978, an employee was required to secure a medical certificate for an absence of less than three days due to illness. This requirement was waived after that date, so that short term absences can be taken without certification, which is still required for longer absences. The results were predictable: absenteeism due to illness for male workers in industry rose from one percent of total annual working days to 1.5 percent by 1981. Short term illness costs are borne by the employer, and they rose from 2.5 percent of hourly earnings in 1976 to 6.6 percent in 1980.[3] As a result of this innovation by the Labor government, any employee can stay home and draw up to three days pay without receiving any medical advice.

Health service cost outlays by the National Insurance scheme as a proportion of GDP are shown in Table VIII-5. They rose by 40 percent from 1970 to 1983 by this measure. Costs of service provided by other public agencies, chiefly municipalities, increased more rapidly, at least up to 1978. Employment in health services, largely in the public domain, grew from 5 percent of total employment in 1970 to 9.2 percent in 1984.

Table VIII-5. Government Supplied Health Service Costs, Norway, 1970-1982

(percent of GDP)

Year	Paid by National Insurance	Paid by all public agencies, including National Insurance
1970	3.8	5.1
1972	4.8	6.1
1974	4.8	7.5
1976	5.4	8.3
1978	4.9	9.1
1980	5.3	n.a.
1982	5.3	n.a.
1983	5.3	n.a.

Sources: Central Bureau of Statistics, Statistical Yearbooks; Stein Kuhnle, Velferdstaten, Oslo, 1983, p. 64.

Although unemployment benefits are generous, costs have not risen rapidly because of the low level of unemployment. In general, the maximum duration of benefits is 40 weeks a year after a 3-day waiting period following registration with the employment service. Daily benefits are set at 0.2 percent of earnings during the previous calendar year. Thus, a person who had earned kr. 85,000 during the year preceding unemployment would be entitled to daily benefits of kr. 700 payable for six days a week, which would come to almost 65 percent of previous earnings. There are supplements for dependent children and vacation pay.

An unemployed person may also secure relocation benefits for moving to a new area. These include start-up costs, family assistance in the event of temporary separation, home visits, and transportation costs of a spouse visiting the new workplace. If the move turns out to be permanent, part of the moving expenses are reimbursable.

Individuals who are older than the normal age of education and who are unemployed by reason of skill deficiencies may be eligible for training benefits. This type of assistance can also be granted if because of age, illness, or other special circumstances, an individual would have little chance of securing employment in his old occupation.

Cash benefits have accounted for only about 60 percent of the cost of unemployment compensation in recent years. Subsidies to firms hiring handicapped workers were 16 percent, followed by training and retraining costs at 8 percent. But the cost of all unemployment programs has been relatively small -- from 2 to 4 percent of social welfare expenditures throughout the 1970s.

The regulations governing occupational safety and health are, if not unique, more stringent than in all but a small number of countries, Sweden among them. In any establishment employing at least five persons, the employees must elect a safety delegate from among their number. In large plants, departmental safety delegates are chosen.

The safety delegate has the right to halt production if in his opinion there is a direct danger to the life or health of employees, and if the danger cannot be prevented without taking such action. Even if this results in economic loss to the employer, the safety delegate is not financially liable. Normally, a safety delegate will not stop production until after consultation with the employer to seek other ways of correcting the conditions causing the complaint. The practice in most countries is to continue work until a government labor inspector has been called in to examine the problem.

A working environment committee must be set up in an enterprise with a minimum of 50 employees. It includes employer and employee representatives and has the function of planning safety and environmental measures. It may call upon the employer to take remedial measures in appropriate cases. The committee must be consulted before any substantial reorganization of operating procedures is undertaken.

There are no meaningful internationally comparative data on the incidence of occupational accidents. In 1980, when 259 such accidents resulted in death in Norway, 65 percent were in oil related activities and shipping, and only 17 in manufacturing.

Children's allowances were paid in 1983 according to the scale shown in Table VIII-6. The Norwegians apparently do not believe that children are cheaper by the dozen.

Table VIII-6. Children's Allowances, Norway, 1983

Number of children under 16 years of age	Annual payment (kroner)	Marginal payment for additional children (kroner)
1	3,816	3,816
2	8,460	4,644
3	14,232	5,772
4	20,436	6,204
5	26,952	6,516
over 5		6,516 each

Source: Rikstrygdeverket, Barnetrygd, January, 1983.

A parent who is single because of the death of a spouse, divorce, separation, or non-marriage, is entitled to a benefit in addition to the entitlement for each child, e.g., the amount for three children if there are in fact two. This tends to equalize family incomes. Allowances are payable regardless of income, and are not taxable, so that they are significant supplements to income.

Conditions of Employment

Prior to 1981, employees had the right to an annual paid vacation of four weeks. The Labor government proposed in 1981 to raise this to five weeks, but the Conservatives modified the entitlement to a gradual increase until 1985, when the full five-week period will come into effect if the state of the economy permits it. Persons over the age of 60 years receive five weeks now, and they will go up to six weeks if a five-week vacation becomes general.

An individual may be hired on a trial basis for a period not exceeding six months. When this period is concluded, six months' notice of discharge is required, with longer notification dependent upon length of service. Notice of discharge must be in writing and for just cause. A breach of the employment contract can lead to immediate discharge, however, though the employee may appeal to the Labor Inspectorate. These provisions do not apply to temporary layoffs.

It was agreed in the 1984 contract settlement that concrete plans for reducing working hours below 40 per week would be considered in 1986. Anyone working on discontinuous shift work or weekends is entitled to a shorter workweek at full weekly pay. Most of these provisions are mandated by law, and are supplemented by collective agreements.

Financing Social Welfare

The welfare programs are financed by payroll taxes supplemented out of general revenues. In both 1970 and 1982, individual employees contributed about one-third of the revenues collected. The employer share rose from 43 to 49 percent between these years, compensated for by a decline in the government contribution and a small change in the revenues derived from other sources.

Rising payroll taxes contributed to higher unit labor costs. In 1974, social security taxes were lowered for both employers and employees in an effort to halt the rising costs, but by 1980, it proved necessary to raise the tax rates once again.

When the pension law was first enacted it was decided that payments should be made out of a central fund, with supplements from general revenues if necessary. The purpose was to accumulate a source of investment funds to replace the loss of private savings that was anticipated because of the new taxes. There was enough in the fund in 1970 to pay 39 percent of the benefits, though the financial plan had called for 71 percent. After 1975, transfers to the central fund virtually ceased, and the idea of building up a pension fund was abandoned. A parliamentary commission explained why:

> That the fund accumulated was less than planned was due both to the strong growth of expenditures and the rise of incomes. The Storting would not raise contributions sufficiently to bring about a further accumulation of funds. In this respect, political

> and economic goals other than national insurance were decisive. . . . Employer and employee contributions were looked upon as instruments of incomes policy and means of improving Norway's international competitive position.[4]

In other words, the prudent intentions of those who founded the national insurance system gave way to the revenue increases necessary to finance an expansionary demand policy during the 1970s. A social security fund might have been a good place for oil revenues, but other uses had priority.

Agricultural Subsidies

It would be difficult to find a country that has not subsidized agriculture in one way or another, but the Norwegian welfare state has been exceedingly generous to its farmers by any standard. The Labor Party by no means bears sole responsibility, since any nonsocialist government must rely for support on the Center Party, which represents farm interests. But Labor has not been backward in bidding for farm votes by means of the agricultural policies it has followed.

Many reasons are advanced in Norway to explain why the government is so solicitous of farmers. There seems to be a nostalgia for the "traditional values" of a rural society. It is also argued that to let farming face free market forces would result in depopulation of rural areas, make Norway too dependent on foreign countries in peacetime, and deprive it of a food reserve in the event of war. Fundamentally, the problem is political. Neither the socialists nor the nonsocialists are prepared to confront the farm bloc. As Odd Aukrust, the former director of research of the Central Bureau of Statistics, put it: "The support system for agriculture is so complicated that few if any individuals have an overall view of the system. . . . As things stand, agricultural policy is an internal matter among the farm organizations, the Department of Agriculture, and the agricultural committee of the Storting. They deal with it in a sovereign manner, and intruders are not welcome."[5]

Agriculture has received larger subsidies than any other sector: 60 percent of total subsidies both in 1973 and 1983. Agricultural subsidies in 1973 equalled 34 percent of the value of farm output, and by 1983, had risen to 89 percent. Output per person employed in agriculture was less than the cost of subsidizing an agricultural job.

The Storting voted in 1975, influenced by prospective oil revenues, to raise the level of farm incomes to that of industrial workers. This goal has been more than achieved, but at considerable cost to the economy. Productive resources have been used less effectively. There has been a loss of consumer welfare as a result of higher food prices, with lower income groups bearing a disproportionate share because of relatively high income elasticity for food. And the subsidy system has become difficult to manage. Egil Bakke, head of the Price Directorate, recently remarked that "the instruments of agricultural policy have become incredibly complicated. It has been said in the past that only Our Lord and the Director of Agriculture fully understand it. Now Our Lord has given up."

Bakke places the blame for the present situation upon a decision adopted some years ago: that Norway should be self-sufficient in grain. Wherever it was possible to raise grain, the government made it profitable through subsidies. Dairy products, vegetables, potatoes, chickens, and eggs gave way to grain. The better farm areas went into grain production, and only marginal farms stayed with other crops. ". . . what we produce in the way of agricultural commodities is produced where conditions are worst (the principle of maximum inefficiency) and nevertheless there is still overproduction."

He questions subsidization of agriculture as a means of stabilizing the existing geographical distribution of population, and concludes:

> Where we could have had a profitable and varied agriculture, we raise grain, which can never be profitable in Norway. The agricultural products that are suited to domestic production are grown on farms which for various reasons can never provide secure work places.[6]

Any real effort to restructure the Norwegian economy will necessarily involve rationalizing agriculture. Many small holdings would have to be abandoned, and the present pattern of crop distribution altered. It would be helpful to the cause of reform if the cost of maintaining the present system were to enter the realm of political debate. The Labor Party is the logical political group to undertake this, since it has the least to lose.

Other Government Programs

Government subsidies have become a way of life in Norway. An overview of the economy written in 1981 contained the following catalogue:

> People who eat bread and drink milk are subsidized; newspaper readers are subsidized; so are people who read books, travel by train, or attend the theater, ballet, or opera. . . . Political parties benefit from government support, as do sporting organizations, music societies, and other cultural activities. Study groups and reading circles are heavily subsidized, as are farmers and fishermen, clothing manufacturers, shipbuilders, and many other enterprises.[7]

Artists, composers, and writers are among those receiving generous state subsidies. For example, many books of poetry are printed in editions of substantial size because all libraries are required to purchase them out of subsidies that they receive from the government for the purpose. Individuals who do not receive support from cultural programs may be eligible for unemployment compensation. Thus far, government programs have not succeeded in producing any modern equivalents of Munch, Ibsen, or Grieg, all of whom had to depend on the demand for their products for their livelihoods.

Industrial Democracy

The rules and regulations that are subsumed under the heading of industrial democracy have become an important aspect of the Scandinavian welfare state. Their effects are difficult to measure quantitatively, but no one doubts that they are among the significant determinants of economic efficiency.

The national agreement between LO and the Norwegian Employers' Association provides for the establishment of a joint labor-management council in any enterprise employing at least 100 persons, as well as in smaller firms if the parties want it. The size of the council varies with the size of the enterprise. Worker representatives are elected by secret ballot, management representatives are appointed by stockholders. The chairmanship rotates between the two groups.

The councils must meet at least once a month. They are entitled to be notified in advance of planned major changes in working conditions, including investment proposals. Matters of health, safety, and training are within their jurisdiction. However, their powers are merely advisory.

Another body, the corporate assembly, was mandated by legislation in 1972. Manufacturing establishments with more than 200 employees must establish these bodies, consisting of at least 12 members, one-third elected by the employees and the rest by the stockholders. All decisions on major investments, reorganization, or changes in production that affect the personnel must be referred to the assembly, which has the right to examine all corporate records. It may adopt resolutions for forwarding to the board of directors. The assembly appoints the board of directors by proportional representation if a third of its members so request, so that in fact, one-third of the board is likely to consist of employee representatives. The system was later extended to construction, newspapers, commerce, transportation, and hotels.

There are mixed opinions about the impact of this Norwegian version of codetermination upon the operation of firms. Few managers feel that the effects are negative, but many believe that they have little influence on the way companies are run.[8] An official of the Employers' Association told the present writer that worker contributions were often useful, but that they preempted a good deal of management time.

Industrial democracy is justified by its proponents as an extension of political democracy to the place of work. It is argued that by giving employees full information about the enterprise, and some influence in managerial decisions, they may make a positive contribution to productivity and are apt to be more reasonable in their demands. They can spur sluggish management in the direction of technological change and restructuring, if the need arises. The other side of the coin is that employees may be reluctant to make concessions if their jobs are at stake.

If maximizing profits ceases to be the objective function of enterprise, the process of policy determination becomes very complicated. It is easy to visualize a situation in which worker representatives on a board of directors might oppose closing a plant, setting up one in another country, or introducing new machinery, even if the accounting data show clearly that these events would augment company profits. The "greater social good" might be served by preserving employment in unprofitable plants as long as payrolls could be met out of assets or through government subsidies. Codetermination could become a powerful force for obstructing structural changes.

There is as yet no convincing analysis of the consequences of industrial democracy in Norway. The German codetermination experience has been studied extensively, but the different traditions of the two countries make it hazardous to extrapolate the findings to Norway. That the system will continue in Norway is beyond question; no political party favors dismantling it.

CHAPTER IX

THE FUTURE OF THE NORWEGIAN ECONOMY

Oil and gas revenues constituted 16 percent of total public revenue and 36 percent of all exports in 1983. The oil sector has become greater than the industry sector, measured as a share of the gross national product. But it is difficult to predict the future contribution of oil to the Norwegian economy because of uncertainties both as to output and price. Production at least is expected to increase, so that cautious optimism is in order. The 1984 National Budget put the matter in proper perspective: ". . . income from oil activities is very uncertain. In order to secure freedom of action in economic policy and to guarantee employment it is necessary to strengthen our traditional economy."[1]

The non-socialist coalition that took over from the Labor Party in 1981 instituted some significant economic policy changes. Unemployment was permitted to rise to levels that Labor would not countenance. Coordinated collective bargaining, in which government fiscal policy and subsidies played a central role, was discontinued, and the labor market partners were allowed to bargain on their own. The role of the state banks in allocating investment resources was reduced and interest rates approached free market levels. Direct control of private bank lending was abandoned. Personal taxes were reduced, and the wealth tax eased.

In general, the present government has tilted economic policy toward a greater degree of market determination and away from regulation by decree. There has also been some modification of the more extreme programs. For example, although self-notification of illness continues, medical certification must be provided sooner as a condition for drawing sickness pay, and there has been a reduction in the number of self-certifications allowed each year. Housing has been deregulated; foreign exchange restrictions have been dropped; several state corporations have been returned to private hands; doctors are now permitted to open private practices and private hospitals are allowed.

It is unlikely, however, that the Norwegian conservatives will attempt to make any serious inroads into the welfare state. Indeed, the 1985 National Budget sets forth the basis for government economic policy in the following terms: "The central goal is to safeguard work for all and to secure and expand the welfare society."

Even the moderate reforms instituted by the Conservatives were attacked by the Labor Party in the 1985 general elections. Labor charged that the national health service was being weakened, and that housing costs had risen substantially. The government was also charged with neglecting some of the more remote areas of the country by failing to let them share in the oil wealth.

The outcome of the elections was extremely close. The Labor Party won 71 seats of the 157 member Storting, and together with the Left Socialists, who also increased their representations, controls 77 seats. The Conservative Party lost 3 seats, and together with its coalition partners, the Christian People's Party and the Center Party, ended up with 78 seats. A small anti-tax group, the Progress Party, managed to hold on to two of the four seats it had held in the previous Storting. As long as the coalition stays unified, the government can remain in office even with a majority of only one because Norway's electoral system does not allow for by-elections or for dissolution of the Storting within the four year period to which its members are elected.

However, the elections reflected the fact that the Norwegian people are not prepared to countenance any substantial reduction in the scope of the welfare state. The election platform of the Labor Party, which is discussed below, called for a further extension of welfare programs, while Prime Minister Kaare Willoch denounced the Progress Party demands for a radical reform of the welfare state as irresponsible. It is clear that while Norwegians are somewhat unhappy about the level of taxes, they like the benefits that the state provides.

The question is often asked: why should the welfare state, at least in its Norwegian version, be criticized? It has achieved many things that other countries would like to emulate. Poverty is a thing of the past, gross disparities in income and wealth have been reduced. Every citizen is guaranteed access to adequate health services. Financial insecurity in old age is no longer a matter of concern. Education is widespread and economically available to all. The hardships of unemployment have been minimized, and the working milieu made as safe as present day technology permits.

Nevertheless, Utopia has not yet been achieved. The welfare state has created problems that may eventually undermine it; a Marxist might say that it contains the seeds of its own destruction. As the welfare state developed in the past, its architects were careful to maintain the economic dynamism that made it possible in the first place. Economic growth was a priority; indeed, the Labor Party had little choice if the ravages of World War II were to be repaired.

As time went on and the memories of austerity faded, other considerations became paramount. Increased consumption, equality of income, and the reduction of working time and effort impinged upon the requirements for economic growth. This trend probably would have continued even in the absence of oil -- it did in the rest of Scandinavia -- but the prospect of oil opened the floodgates. The once prudent Norwegians threw caution to the winds and incurred a heavy foreign debt to finance the mature welfare state in which the pattern of life was determined increasingly by the government.

The Work Ethic

The willingness to work has been impaired by the welfare state. It has become increasingly difficult to recruit people for routine jobs, not only because the labor market is tight, but also because alternatives to work are available.

This is difficult to document statistically in the absence of suitable data. For example, one would expect that the ratio of job vacancies to job seekers would tend to increase over time as the unemployment became more selective. Employment office data indicate that this ratio has varied over the business cycle, but shows little time trend. However, job vacancy data are notoriously difficult to interpret.

An analysis of attitudes toward work based on a 1976 survey revealed that of 800,000 people of working age outside the labor market, some 200,000 to 240,000 wanted paid work but were not looking for it because they felt that no suitable jobs were available.[2] This so-called discouraged worker effect is a common phenomenon in countries with substantial unemployment. But with unemployment at 1.8 percent in 1976, it is difficult to escape the conclusion that many Norwegians had set up very high job suitability requirements and were easily discouraged from entering the labor market because they had other means of support.

One additional set of statistics might be cited in support of the hypothesis that the propensity to work has been impaired. From 1975 to 1983, the average annual number of working days lost due to illness rose from 29 to 34 for all males. For men aged 16 to 29 years, the increase was from 14 to 22 days; aged 30 to 49, 19 to 26 days; and aged 50 to 66, 43 to 46 days. Similar behavior was recorded for women.[3] This suggests that younger people, who were perhaps not imbued with a strong work ethic, were taking greater advantage of the looser self-reporting requirement introduced in 1978.

High income tax rates also tend to reduce work incentives. The ratio of overtime hours to total hours worked by adult males in manufacturing fell from 4.4 percent in 1970 to 3.8 percent in 1984 despite the fact that working hours had fallen. High marginal taxes also tend to produce a "black" labor market. In a report to the Storting in 1981, the Labor government conceded that the "black" economy was becoming a problem: "In Norway there are many indications of an extensive and increasing unregistered and illegal economic activity. . . . While it used to be considered reprehensible, there is now a tendency for more people to participate in it if they can. . . . Younger people are believed to react less negatively to the black economy than older people. . . . The government feels the black economy is very worrisome. A growing black economy is a sign of a lack of solidarity. Through such activity some people avoid paying their share of the common expenditures in society." It is interesting to note that the government's proposed remedies were stronger controls: "The welfare state is not conceivable without large common expenditures and extensive transfers to vulnerable groups. <u>This means that the level of the gross taxes cannot be reduced</u>."[4] [Emphasis added.]

Work motivation is difficult to document with any degree of precision. People are not likely to admit to an unwillingness to work as long as it is regarded as a socially desirable activity. Moreover, as a society becomes wealthier, there is a long run tendency to substitute leisure for work. However, it seems likely that the combined effects of generous support of non-work plus financial penalties for a high level of work activity have reduced the willingness of Norwegians to engage in regular paid employment.

Entrepreneurial Activity

A good deal of the dynamism of capitalist economies

stems from the efforts of private investors in quest of profits earned through the introduction and marketing of new products. For example, much of the new electronic technology in the United States originated in "silicon valley" firms, founded mainly by young engineers. Such activity is less likely when rates of return on capital are low. The data in Table IX-1 suggest that this has been a problem in Norway. It may also be true, as Noreng argued, that "the Norwegian desire for stability is insufficiently in accord with the capitalist requirement of movement and instability."[5]

Table IX-1. Rates of Return on Capital[a] in Selected OECD Countries, 1970-1978

	1970	1972	1974	1976	1978
Norway	8.6	8.0	11.1	7.2	3.8
Sweden	----	----	11	----	----
United States	21	25	18	23	24
Japan	39	27	26	24	----
Germany	21	16	16	16	----
United Kingdom	10	9	3	3	5

Source: Report No. 79 to the Storting (1980-81), Norwegian Long-Term Programme, 1982-1985, p. 82.

[a]Defined as the ratio between operating profits at current prices less compensation for owner services to fixed real capital at current replacement costs.

The manner in which the oil discoveries have been exploited tended to strengthen the role of government in the economy. It has been made easier to control the allocation of investment resources. The stock market has not been a major source of risk capital, as it is in many countries. In 1983, only 6 percent of all domestic capital was raised in this way. The maintenance of low interest rates did not allow sufficient flexibility for different degrees of risk.

Government financing tended to crowd out private capital, and investment was often based on political rather than economic criteria. Too much control of investment was vested in public officials with no personal stake in the outcome. The Labor government's own department of industry said as much:

> . . . with the emphasis on a dynamic industrial structure, it is necessary to have a sufficient number of new enterprises. . . . [Various studies] indicated that there may have been too few new enterprises in relation to what is desirable. There may be several reasons. Among others, it may be that economic returns do not stand in reasonable relationship to the risk involved in setting up a new firm. Another major reason may be the many rules and regulations involved in starting a new handicraft or manufacturing establishment.[6]

The Labor Market

The economic policies of the 1970s all but destroyed the labor market as a mechanism for allocating manpower in an efficient manner. The frictional unemployment that is essential if workers are to be employed where their productivity and incomes are highest was kept unduly low by the maintenance of very high demand levels. An adequate supply of working skills was endangered by the narrowing of wage differentials to what must be among the smallest in the world. The orderly determination of wages was complicated by a bargaining system in which there was an open-ended invitation to firms that could pass increased labor costs to the consumer and to the government to buy labor peace without putting up much resistance.

Any centralized system of collective bargaining is bound to lead to some wage drift to accommodate local requirements to national norms. In the past, the LO and the Employers' Association were able to exercise sufficient discipline over their members to prevent wage drift from becoming excessive. During the 1960s, wage drift contributed half of the average annual increase in hourly earnings. From 1977 to 1982, the wage drift share had risen to 70 percent, and became the principal mechanism for determining wages.

The Labor Party and LO do not bear the exclusive responsibility for this development. Stronger companies liked the system. Central negotiations in 1984 led to an average wage increase of 5.9 percent. Aftenposten, the country's leading newspaper, added another 4 percent for its employees. Other less profitable papers had to follow along. Indeed, many of these papers would have gone out of existence were it not for government subsidies.

Changing established ways of doing things is never easy, and it is hard to say whether the current debate in Norway will have any effect on the collective bargaining system. Preben Munthe, for many years the principal government labor mediator, argues that neither a completely centralized nor decentralized bargaining model will work. For him, the best way out is to educate people so they understand that money wage increases in excess of productivity are bound to be inflationary and to no one's benefit.[7] Munthe is a university professor, which may explain his faith in education.

Hermod Skånland, who served as chairman of the committee that recommended revamping the bargaining system a decade ago, is of the view that progress can only be made toward a less inflation-prone collective bargaining process if it is accepted that under present conditions, the government cannot guarantee full employment. This would oblige the bargaining parties to consider more seriously the employment consequences of their demands and concessions. He retains a belief in the efficacy of centralized bargaining, but only if those who have responsibility also have the power to make decisions. Income distribution and goals other than employment should be left to taxes and subsidies.[8]

Egil Bakke, the price director, poses four questions that he believes to be relevant in approaching reform of the bargaining system:

1. To what extent should employer and employee "monopolies" -- employer associations and trade unions -- have equal rights with labor and management in individual enterprises to bargain and to declare work stoppages?

2. To what extent should those employed in monopoly enterprises have the same right to bargain and strike as those in enterprises subject to competition?

3. To what extent should the general rules for negotiation apply to enterprises and industries that derive a significant portion of their funds from the state budget?

4. Are the general negotiation rules and procedures rational for employees of the central government and

municipalities?

He offers no answers beyond saying that the present rules are "irrational, unreasonable, and dangerous."[9]

There is nothing wrong in principle with letting each enterprise fix its own wages, depending on productivity and profitability. But this presupposes the existence of a regimen in which the consequences of excessive liberality are borne by the firm and its employees. If unprofitable firms must pay the same wages as profitable ones, particularly if the latter are protected from competition and the government is prepared to underwrite costs for the former group, then Preben Munthe is correct in asserting that no one benefits, and inflation will be the result. Given the quasi-monopolies that participate in the Norwegian bargaining system, any institutional arrangement is likely to fail unless the rules of the game are changed to oblige the parties to face bankruptcy or unemployment, as the case may be.

In sum, the question is whether the Norwegian welfare state is consistent with traditional collective bargaining in which independent trade unions deal with private employers. The answer is probably no, particularly when there is a socialist government intent on combining full employment, equality, and economic security for all with a satisfactory rate of economic growth. Government intervention then becomes inevitable unless Norwegians are educated to a level of sophistication that does not yet prevail anywhere else.

The Transfer Economy

This is an alternative way of describing the welfare state. What it means is that a large proportion of the goods and services consumed by households is supplied by government. The original intent, in setting up transfer programs, was to redistribute income from the wealthy to the poor. As things have worked out, a substantial portion of the transfers are distributed without regard to need. Indeed, the redistribution effect may be negative, as in the case of subsidized home mortgages.

The bewildering complexity of individuals, organizations, occupational groups, and enterprises that have become clients of the welfare state makes it difficult to determine the net effect on social welfare. Does society gain if the government supports activities, in the interest of full employment, that result in the production of goods and services for which there is no effective demand? What are the criteria for determining who is entitled to remain in his or her present line of work with government subsidy, and who should be obliged to seek a new occupation? Norway is still a long distance from the Soviet system under which art, music, and literature have been stifled by the "unions" of these professions, which decide to whom government support should go, but it is somewhat disquieting when a Labor government's long-term program proclaims that it "will consider whether the state should assume an even greater responsibility for the working conditions of artists in society."[10]

It is clear that the welfare state is not without its blemishes. All the recipients of transfer payments are prepared to accept their entitlements, but not all are willing to pay for them through high taxes. A large number of public servants are required to transfer an individual's income from one pocket to another. The ability of people to spend their incomes according to their own tastes is reduced.

A pernicious practice of the United States Congress that makes it difficult to control government expenditures is called "logrolling." A representative of a state or congressional district who wants a project for his area -- a bridge, a road, a dam, a military camp -- trades his vote with other representatives who are seeking similar projects for their constituents. The result is that many structures of dubious economic merit are installed. The welfare state simply enables this to be done on a national basis, with the trading done by political parties on behalf of large interest groups.

Economic Growth and Profligacy

At first blush, Norway's recent record of economic growth looks quite satisfactory. From 1976 to 1983, inclusive, GDP growth in Norway averaged 3.2 percent per annum, compared with 2.3 percent for the OECD as a whole. However, if the oil sector is omitted, the Norwegian growth figure falls to 2.2 percent.

Potential growth of the non-oil economy was sacrificed to finance consumption. The OECD, which always hesitates to criticize a member nation, put it rather cautiously:

> Norway is one of the few countries (together with Sweden and the United Kingdom) where the present [1982] level of production of manufactures is lower than the 1973/1974 peak. Overall, traditional sectors would probably have fared better if the degree of absorption of oil revenues had been smaller and the pattern different, in particular if there had been room for tax cuts as an alternative to higher public expenditures.[11]

If economic growth is an objective for Norway, the capital derived from oil should be invested rather than consumed. This was the intent, at least in part, of the 1974 Labor government report on the future role of oil in the Norwegian economy. While some of the proceeds were to be utilized for raising domestic consumption, it was stressed that since the oil industry would entail a reconstruction of Norwegian industry, consumption growth "should be limited and extended over a longer period of time. . . . A considerable proportion of the public revenues from petroleum operations must be invested abroad."[12]

This was a good set of guidelines, but it was not followed. Had it been, the Norwegian economy would have looked quite different today. If industrial restructuring had been encouraged, the additional investment resources made available might well have resulted in a product mix better suited to both domestic and foreign demand. Instead of paying off foreign creditors, as it has done for the last five years, Norway would have begun to reap the benefits of foreign investments.

When an individual inherits a fortune, there is a great temptation to raise his living standards immediately. A prudent person will do so in moderation, with an eye toward his future and that of his children. Capital will be conserved and the increase in consumption will come out of income from the capital.

A profligate individual will throw caution to the winds and indulge himself in luxuries. In the end, he may be worse off than when the inheritance was received. Norway is certainly not an example of a profligate state, but neither can its behavior over the past decade be called prudent.

Lack of knowledge was certainly not the problem. There were many in government who knew what ought to be done, but the politicians did not listen to them. One wonders why so little attention was paid to the repeated warnings that appeared, *inter alia*, in the publications of the Bank of Norway. Perhas it is inevitable that politics should be in command in a democracy, but the particular brand of politics that controlled Norway in the 1970s proved to be unable to generate a satisfactory economic performance.

What Has Labor Learned?

The Norwegian Labor Party increased its share of the popular vote from 37.2 percent in 1981 to 40.8 percent in 1985. If the votes won by the Left Socialists are added, it would apear that not far from half of the Norwegian electorate favors some form of socialism. While it is unlikely that Labor will come to power before 1989 at the earliest, the Party's 71 Storting members will probably continue to oppose any attempts to dismantle the welfare state that they created. The question of what the Labor Party and its allied trade unions have learned from the experience of their last term in office is not purely academic.

The question is also of more general interest for the light it may throw on the future of European social democracy. We know that the communist model of government-owned industry does not perform as efficiently as the relatively pure capitalist model exemplified, for example, by the United States, Germany, and Japan. What can we expect of the half way house, "the middle way," represented by social democracy, in which enterprise is still largely in private hands but where the government plays a major role in the allocation of resources through monetary and fiscal policy as well as direct intervention?

In preparation for the 1985 general election, the Labor Party issued a 100-page book setting forth its views on domestic and foreign policy.[13] Entitled *New Growth for Norway*, it represents the current thinking of the party leadership. In Norway, a political program is not mere campaign rhetoric. If Labor had won, there would have been strong pressures to oblige it to live up to its commitments.

Basic Policy Goals

The program is prefaced by a commitment to restore

full employment and to expand the welfare state. The third leg of the tripod, greater equality, does not appear right at the start, although there are many references to it elsewhere.

The emphasis throughout is on collective, as opposed to individual, responsibility and action. "It is necessary to guarantee adequate collective consumption. Collective consumption will receive priority over any increase in private purchasing power."

Labor's 1981 manifesto acknowledged that there are limits to the welfare tasks that the government can undertake, and anticipated that there would be supplementary efforts by such smaller social groups as the family, fellow employees, and the churches. The new program appears to move in the opposite direction: "The tasks of a welfare society cannot be solved through privatization. Transfer of more tasks to the private sector, to the family, and increased resort to individual co-sharing of [medical and other] fees lead to greater inequality, increases in class differences, and conflicts between individuals and groups." The welfare state must be based on "solidarity, comradeship, and mutual responsibility."

Oil

It is proposed that the output of oil be expanded gradually, although there are no specific targets. However, there would be more centralized control of the industry: "Statoil is to be built up further as an integrated company. Political direction will be guaranteed by strengthening the Department of Oil and Energy and by submitting decisions on oil to the Storting." Participation by international oil companies would be reduced.

Part of the oil revenues would go into a state development fund for smoothing fluctuations in the business cycle, stimulating the formation of new firms, assisting in industrial reconstruction, and promoting foreign investment. The mechanisms by which these objectives would be achieved are not specified, but presumably the directing agent would be the government.

Prices and Incomes

Prices would be subject to strict control where there is insufficient competition. Income differentials not based on particularly onerous, responsible, monotonous, or dangerous work are to be opposed. Wealth, particularly when it is inherited, perpetuates inequitable social distinctions. This problem would be handled in part by making income taxes more progressive.

There is no commitment to a reduction in the overall tax burden, merely a redistribution of the incidence. Tax reform is expected to lead to "strengthening collective consumption, to a just distribution of income, and to a better direction of social resources to growth, production, and employment."

How future inflation of wages and prices is to be prevented is largely ignored. Wage determination would be left to the labor market organizations, with the government cooperating through appropriate employment and distribution policies. However, small union and employer groups would not be permitted to break out of a centrally determined incomes policy. The only recognition of the wage drift problem is the statement that "moderation on the part of employees can result in too great an increase of ownership income in enterprises with substantial ability to pay wages. This will be counteracted by tax and price policies." Does this mean that the more profitable firms are to be subjected to price control and higher taxes, so that they are better able to resist importunate wage demands?

Employment Conditions

Manpower forecasts anticipate that there will be a need for between 150,000 and 200,000 new jobs between 1984 and 1990 if unemployment is not to rise. The Labor Party program proposes new measures to maintain full employment. Government employment at normal wage rates would be guaranteed to everyone under 20 years of age who is registered as unemployed, and consideration would be given to the extension of this guarantee to the 20-25 year age group should there be any substantial unemployment among them.

This proposal raises an interesting question. Would unemployed youths be obliged to accept government jobs or face the termination of their unemployment benefits? The purpose of benefits is to provide income maintenance while an unemployed person is searching for suitable employment. This opportunity, which is particularly important for young people who have not yet determined their career goals, might be short circuited by obligatory government employment, if that were the intent.

The Labor Party would reduce the costs of labor mobility for those who are disposed to move. But the existing population pattern is to be preserved, so that jobs are to be provided where there is unemployment, rather than have the pressures of unemployment force people to move.

It is of interest to note that the coolness of the Labor Party to labor mobility is not shared by the present conservative government. Mobility has been furthered by requiring recipients of unemployment benefits to accept jobs entailing one hour of commuting time each way, with even longer travel a possibility except for families with small children where both parents are employed.[14]

Other labor market initiatives proposed by the Labor Party's program include paid study leaves for adults and new training courses. Compulsory quotas of 40 percent for women would be established in areas of work where they are under-represented, and special stipends given to women who decide to enroll in courses of study traditionally limited to men. Enterprises could receive subsidies if they employed women in normally male jobs. A long-term goal of a six-hour day is set forth, with a 37½-hour week to be supported in the short run.

Government Income and Expenditures

Labor's 1981 election program stated that it was "absolutely necessary to reduce the sharp growth in public sector expenditures." This is not repeated in 1985. On the contrary, since industrial employment is not likely to expand, it is the public sector that must be the engine of economic growth. Local government in particular must provide additional services, financed by central government grants. National plans for employment in sports and cultural activities are also on the agenda.

If Labor's ambitious schemes for expanding the state sector were realized, Norway would probably become the non-communist world's leader in government spending. How is all this to be financed? By higher taxes, domestic or foreign borrowing, oil revenues? The program is silent on this question.

Industry

A recent joint study by the trade unions and the

manufacturers' association begins with the observation that "the development of traditional industrial activity has been weak during the past decade. Adjusted for inflation, value added and output of goods-producing industry are lower now than in the mid-1970s."[15] A Labor Party document devoted to industrial policy contained some specific recommendations for revitalizing industry.[16] By contrast, the 1985 program is couched in vague terms. Its main elements are:

1. "Government authorities must play an active role in strengthening industrial possibilities both through general economic policy and by direct initiatives. . . The authorities will be responsible for national planning and for developing a strategy for industrial development. Analysis of important branches of industry and the various geographical areas will provide signals for the types of industry to be proposed."

2. Although some risk capital will be provided by the stock market, "the government must contribute to industrial financing both by providing adequate credit and by direct participation." However, care must be taken lest tax regulations permit "unreasonably low taxation of non-labor income or lead to investment that is purely tax motivated."

3. An oil revenue fund will support industrial research, development, and restructuring.

4. The purchasing power of the government itself and of such governmental sectors as communications, transportation, water power construction, and defense are to help Norwegian industry. (But what would be the impact on exports if other governments adopted similar autarchic policies?)

Other commitments are scattered throughout the document. A national telecommunications network is to be established. There will be heavier investment in roads and railroads. Shipping will be supported in various ways. Running through the broad agenda is the key role of government. "It is important to find means of implementation and a decision process that will ensure that bodies elected by the people have adequate influence over industrial policy."

The Labor Party clearly does not rely on private entrepreneurs to make the right investment decisions, and is not prepared to let the traditional capitalist incentive of profit maximization steer industrial development.

Banking and Credit

What the program envisions under this head is a return to the policies of the 1970s. Interest rates are to be kept at relatively low levels. The state banks will play a central role in the provision of credit, particularly in home building. To ensure that private bank lending decisions are not made on narrow commercial grounds, their managing bodies will consist of "representatives of government, employees, and stockholders, with no group having a majority." The banks would thus be "redemocratized," reversing the present government's alteration of Labor's 1978 statute. The private banks have become a political football.

Industrial Democracy

It is the intention of the Labor Party to extend the scope of industrial democracy in order to provide "greater equity and power sharing in economic life." This will, it is hoped, improve the quality of management and sharpen enterprise responsibility in meeting social obligations in matters of employment, health, and environment. Employee representation in management bodies would be increased; all types of firms, not only corporations, would be covered; and smaller firms, thus far exempted, would also be required to provide for employee participation in management.

The Welfare State

The 1985 Labor program envisions a large expansion of state welfare activities. To name a few:

1. Better conditions for artists.

2. Assistance to independent theater and dance groups.

3. A comprehensive plan for Norwegian musical life.

4. Higher subsidies to libraries.

5. More suport to the publication of literary and vocational books.

6. Additional government subsidies to newspapers

in order to ensure freedom of speech and the presentation of different points of view, particularly to preserve competitive papers in larger towns.

7. Increased government outlays for sports installations, camping, hunting and fishing facilities, small boat marinas.

8. More state-supported nurseries for children.

9. Recreational centers, special transportation, and home assistance for the aged.

10. Expansion of the health network, particularly in primary health services and psychiatric treatment.

11. Maternity leave with pay would be extended from the present 108 days to 30 weeks, plus two weeks for the father.

Agriculture is handled in a cautious fashion. Its importance to the economy is acknowledged. There is need for further income equalization <u>within</u> the farm sector to the benefit of small and less well endowed farms. To eliminate overproduction, part-time employment would be provided. There is no mention of any desirable change in the current subsidy system.

<u>A Final Word</u>

In setting forth its 1985 election manifesto, the Labor Party does not appear to have benefited from experience. It should be clear by now that government control of the capital market is not likely to produce an optimum allocation of investment resources and thus to maximize growth. Extension of the welfare state may further erode individual initiative, as will concentration on equality of income rather than equality of opportunity. Government subsidies to industry and agriculture do not promote their efficient operation. Unemployment that approaches the zero level is attainable only at high cost to the economy. Inflation cannot be prevented by direct government controls except in the short run, and then only by distorting the economy.

None of these things is reflected in recent Norwegian socialist thinking. On the contrary, there appears to be more emphasis than ever on collectivism, on state direction of economic and cultural life. There is insufficient evaluation of the consequences that may flow from this step to the left.

This is not to say that a socialist Norway would necessarily fall into an economic slough. There is always oil to bail out erring governments, as happened in the 1970s. The danger is that the mentality that made pre-oil Norway affluent, the drive that Norwegian migrants brought with them to the United States and led to their success, will be replaced by a more passive rentier philosophy.

The Labor Party has suffered a great deal of internal turmoil in recent years, and a new generation has come to power. It remains to be seen whether pressures from the radical left for further redistribution of income and wealth, and for collective rather than individual action, will be resisted. The 1985 manifesto is not promising. If it reflects what Norway will be like under a future Labor government, serious questions that go beyond Norwegian borders are raised. In other countries of Northern Europe -- Great Britain, Germany, Denmark, Sweden, the Netherlands -- social democratic parties have veered to the left in the last five years. Many who view this trend with apprehension, including those who have been sympathetic with the aspirations of traditional social democracy, are beginning to ask whether liberal economic institutions can survive socialist hegemony. Does the logic of socialism lead inevitably to greater and greater centralization of power in the state?

The danger is that if this trend continues, socialists, when they attain office, will use the formidable levers of government to install centralized patterns of social organization that will be difficult to reverse even when they lose favor with the electorate. Once established, the socialist welfare state seems to acquire a life of its own. There may be no turning back toward a society that offers the individual less protection from economic vicissitudes but more opportunity for personal initiative.

During the next four years, the non-socialist coalition that retained office by a slender margin may succeed in moving the country toward greater reliance on free market forces as the principal mechanism for allocating resources. The socialists will have more time to consider possible alternatives to their present views on how the economy should operate. And the Norwegian people will have an additional opportunity to determine which system is more likely to produce the "new growth for Norway" that Labor proclaimed as its central goal in the 1985 election campaign.

NOTES

CHAPTER I: THE NORWEGIAN ECONOMY ON THE EVE OF THE OIL DISCOVERIES

1. See OECD, National Accounts 1952-1981, p. 86. This ranking was measured in U.S. dollars at the exchange rates and price levels prevailing in 1975. Alternative measures would not greatly affect the ranking.

2. OECD Economic Surveys, Norway, 1973, p. 5.

3. Robert J. Flanagan, David W. Soskice, and Lloyd Ulman, Unionism, Economic Stabilization, and Incomes Policies, Washington, 1983, p. 160.

4. OECD Economic Surveys, Norway, January 1971, p. 44.

CHAPTER II: OIL

1. The data are from Norges Offentlige Utredninger (hereafter abbreviated as NOU), Petroleumsvirksomhetens framtid, 1983, No. 27, p. 17.

2. Norwegian Central Bureau of Statistics, Economic Survey, 1981, p. 129.

3. OECD Economic Surveys, Norway, 1980, p. 43.

4. Petroleumsvirksomhetens framtid, op. cit., p. 46.

5. Ministry of Finance, Petroleum Industry in Norwegian Society, Parliamentary Report No. 25, 1973-74, p. 9.

6. Ministry of Petroleum and Energy, Relating to Perspectives in Petroleum Activity in the Coming Years, Report No. 40, 1982-83, p. 36.

7. Petroplan A.S., "Managing the Petroleum Sector," in Oslo Gruppen, Full Sysselsetting og Okonomisk Vekst uten Inflasjon, Oslo, 1981, Appendix 5.

8. Ministry of Finance, Petroleum Industry in Norwegian Society, op. cit.; Natural Resources and Economic Development, Parliamentary Report No. 50, 1974-75.

9. Ministry of Finance, Petroleum Industry in Norwegian Society, op. cit., p. 53.

10. Norges Bank, *Economic Bulletin*, 1974, No. 1, p. 17; 1974, No. 4, pp. 172-160.

11. *Ibid*., 1975, No. 4, p. 170.

CHAPTER III. LABOR'S POLICY OBJECTIVES

1. Bent Roald Andersen, "Rationality and Irrationality of the Nordic Welfare States," *Daedalus*, Winter, 1984, p. 118.

2. The internal divisions within the Labor Party are analyzed in Haakon Lie, *Slik jeg ser det*, *No. 2*, Oslo, 1983.

3. Ministry of Finance, *Natural Resources and Economic Development*, *op. cit*., p. 49.

4. The National Budget of Norway, 1980, p. 20.

5. Norges Bank, *Economic Bulletin*, 1981, No. 4, p. 294.

6. Ministry of Finance, *Natural Resources and Economic Development*, *op. cit*., p. 50.

7. Norges Bank, *Economic Bulletin*, 1982, No. 4, p. 248.

8. Ministry of Finance, *Natural Resources and Economic Development*, *op.cit*., p. 40.

9. Hans F. Dahl, "Those Equal Folk," *Daedulus*, Winter, 1984, p. 97.

10. Norges Bank, *Economic Bulletin*, 1977, No. 4, p. 15.

11. OECD Economic Surveys, *Norway*, 1975, p. 28.

12. Norges Bank, *Economic Bulletin*, 1977, No. 4, p. 221.

13. See OECD *Historical Statistics*, 1960-81, p. 102.

14. The National Budget of Norway, 1972, pp. 6-7.

15. *Ibid*., 1974, p. 9.

16. *Ibid*., 1977, p. 3.

17. *Ibid*., 1978, p. 18.

18. *Ibid.*, 1979, pp. 20-21.

19. *Ibid.*, 1981, pp. 22-23.

CHAPTER IV: WAGES AND EARNINGS

1. Flanagan, Soskice, and Ulman, *op. cit.*, p. 173.

2. Norges Bank, *Economic Bulletin*, 1978, No. 3, p. 181.

3. OECD Economic Surveys, *Norway*, 1976, p. 13.

4. *Ibid.*, 1978, p. 30.

5. Norges Bank, *Economic Bulletin*, 1978, No. 4, p. 246.

6. Norske Bankforening, *Okonomisk Revy*, 1978 No 1, p. 5.

7. *Ibid.*, 1982, No. 3, p. 9.

8. Speech of Tor Halvorsen before the Representative Council of LO, June 7, 1983 (mimeographed).

9. Tor Halvorsen, *op. cit.*

10. OECD Economic Surveys, *Norway*, 1971, p. 39.

11. *Ibid.*, 1982, p. 42.

12. *Ibid.*, 1983, p. 34.

CHAPTER V: EMPLOYMENT AND UNEMPLOYMENT

1. The data are from Central Bureau of Statistics, *Labor Market Statistics*, 1973 and 1974; Ministry of Petroleum and Energy, Report No. 32 to the Storting, 1984-85, p. 56.

2. NOU, *Strukturproblemer og vekstmuligheter i Norsk industri*, 1979: 35, 73.

3. See St. meld. No. 85, *Om Sysselsettingen og arbeidsmarkedspolitikken*, 1982-83, pp. 87-93.

CHAPTER VI: MANAGING THE ECONOMY

1. The data are from the Revised National Budget of Norway, 1985, p. 3.

2. OECD Economic Surveys, *Norway*, 1983, p. 37.

3. Norges Bank, *Economic Bulletin*, 1974 No. 4, p. 153.

4. *Ibid.*, 1980 No. 4, p. 244.

5. *Ibid.*, 1971 No. 4, p. 125.

6. National Budget of Norway (English Summary), 1980, p. 33.

7. *Ibid.*, 1983, pp. 29-34.

8. Norges Bank, *Economic Bulletin*, 1982 No. 4, p. 249.

9. Revised National Budget of Norway, 1984-85, p. 20.

10. National Budget of Norway (English summary), 1980, p. 36.

11. Norwegian Commercial Banks, *Financial Review*, June 1979, p. 5.

12. OECD Economic Surveys, *Norway*, 1981, p. 32.

13. *Okonomisk Revy*, No. 2, 1980, pp. 4-6.

14. *Ibid.*, No. 5, 1980, p. 14.

15. Norges Bank, *Economic Bulletin*, 1980 No. 4, p. 257.

16. *Ibid.*, 1974 No. 2, pp. 67-70.

17. *Ibid.*, 1976 No. 4, p. 213.

18. Norwegian Commercial banks, *Financial Review*, 1977, No. 1, pp. 1-8.

19. *Ibid.*, 1977, No. 4, pp. 1-3.

20. OECD Economic Surveys, *Norway*, 1980, p. 17.

21. National Budget of Norway (English summary), 1983, p. 45.

22. *Ibid.*, 1982, p. 24.

23. Norges Bank, *Economic Bulletin*, 1972 No. 3, p. 136.

24. *Ibid.*, 1974 No. 4, pp. 154-157.

25. OECD Economic Surveys, *Norway*, 1977, pp. 5-10.

26. *Sosialokonomen*, 1981, No. 5, pp. 11-12.

CHAPTER VII: INDUSTRY AND FOREIGN TRADE

1. Oystein Noreng, *Olje-Norge*, Oslo, 1984, p. 43.

2. Central Bureau of Statistics, *Historical Statistics*, 1978, pp. 214-215.

3. Landsorganisasjonen og industriforbund, *Norge som industrinasjon*, 1984, p. 6.

4. NOU, *Strukturproblemer og vekstmuligheter i Norsk industri*, 1979, No. 35, p. 56.

5. OECD *Historical Statistics*, 1960-1982, p. 46.

6. OECD Economic Surveys, *Norway*, 1983, p. 27.

7. Department of Industry, *Industripolitiske Retningslinjer*, St. meld. No. 54, 1980-81, p. 14.

8. Memorandum to Prime Minister Odvar Nordli, June 4, 1979 (mimeographed).

9. *Ibid.*

10. Norwegian Commercial Banks, *Financial Review*, 1978, No. 1, p. 8.

11. Speech by Tor Halvorsen to the Industry Forum, March 13, 1982 (mimeographed).

12. It may be noted, however, that closing Tandberg did not give rise to any appreciable unemployment and that creditors were fully paid.

13. Norges Bank, *Economic Bulletin*, 1978 No. 4, p. 245.

14. OECD Economic Surveys, *Norway*, 1979, p. 25.

15. Knut Getz Wold in Norges Bank, Economic Bulletin, 1981 No. 1, p. 7.

16. Oystein Noreng, op. cit., p. 40.

17. Ibid., p. 61.

18. OECD Economic Surveys, Norway, 1983, p. 26.

19. For an excellent review of government policy, see Hermod Skanland, "Norwegian Exchange Rate Policy" Norges Bank, Economic Bulletin, 1983 No. 1, p. 56.

20. Strukturproblemer og vekstmuligheter i Norsk industri, op. cit., p. 60.

21. Oystein Noreng, op. cit., p. 134.

22. Ibid., p. 77.

CHAPTER VIII: THE WELFARE STATE

1. NOU 1984: 10, Trygdefinansiering, pp. 82-83.

2. Trygdekontoret, Utgifter som folketrygden dekker ved sykdom, August, 1981.

3. OECD Economic Surveys, Norway, 1982, pp. 34-40.

4. NOU 1984:10, op. cit., p. 34.

5. Nationen, August 4, 1984.

6. Aftenposten, March 8, 1984.

7. Oslo Gruppen, op. cit., Appendix I, pp. 13-14.

8. International Labor Organization, Workers' Participation in Decisions Within Undertakings, Geneva, 1981, pp. 95-97.

CHAPTER IX: THE FUTURE OF THE NORWEGIAN ECONOMY

1. Revised National Budget of Norway, 1984, p. 18.

2. Olaf Foss, *Attitudes and Behaviour in the Labor Market*, Oslo, 1980.

3. Statistical Yearbook of Norway, 1985, Table 71.

4. Report No. 79 to the Storting, 1980-81, *Norwegian Long Term Programme 1982-85*, p. 210.

5. Oystein Noreng, *op. cit.*, p. 60.

6. *Industripolitiske retningslinjer*, *op. cit.*, p. 87.

7. *Aftenposten*, June 9, 1984.

8. *Ibid.*, September 7, 1984.

9. *Ibid.*, June 14, 1984.

10. *Norwegian Long Term Programme*, *op. cit.*, p. 185.

11. OECD Economic Surveys, *Norway*, 1982, p. 28.

12. Report to the Storting No. 25, *Petroleum Industry in Norwegian Society*, 1973-74, p. 18.

13. Arbeiderpartiet, *Ny Vekst for Norge*, 1985.

14. *Om sysselsettingen og arbeidsmarkedpolitikken*, St. meld. No. 85, 1982-83, p. 74.

15. Landsorganisasjonen og Norges Industriforbund, *Norge som Industrinasjon*, 1984, p. 1.

16. *En Handlingsplan for Norsk Industri*, August 23, 1983 (mimeographed).

INDEX